FTTX Concepts
and Applications

FTTX Concepts and Applications

Gerd Keiser
PhotonicsComm Solutions, Inc.

IEEE PRESS

A JOHN WILEY & SONS, INC., PUBLICATION

Copyright © 2006 by John Wiley & Sons, Inc. All rights reserved

Published by John Wiley & Sons, Inc., Hoboken, New Jersey
Published simultaneously in Canada

No part of this publication may be reproduced, stored in a retrieval system, or transmitted in any form or by any means, electronic, mechanical, photocopying, recording, scanning, or otherwise, except as permitted under Section 107 or 108 of the 1976 United States Copyright Act, without either the prior written permission of the Publisher, or authorization through payment of the appropriate per-copy fee to the Copyright Clearance Center, Inc., 222 Rosewood Drive, Danvers, MA 01923, (978) 750-8400, fax (978) 750-4470, or on the web at www.copyright.com. Requests to the Publisher for permission should be addressed to the Permissions Department, John Wiley & Sons, Inc., 111 River Street, Hoboken, NJ 07030, (201) 748-6011, fax (201) 748-6008, or online at http://www.wiley.com/go/permission.

Limit of Liability/Disclaimer of Warranty: While the publisher and author have used their best efforts in preparing this book, they make no representations or warranties with respect to the accuracy or completeness of the contents of this book and specifically disclaim any implied warranties of merchantability or fitness for a particular purpose. No warranty may be created or extended by sales representatives or written sales materials. The advice and strategies contained herein may not be suitable for your situation. You should consult with a professional where appropriate. Neither the publisher nor author shall be liable for any loss of profit or any other commercial damages, including but not limited to special, incidental, consequential, or other damages.

For general information on our other products and services or for technical support, please contact our Customer Care Department within the United States at (800) 762-2974, outside the United States at (317) 572-3993 or fax (317) 572-4002.

Wiley also publishes its books in a variety of electronic formats. Some content that appears in print may not be available in electronic formats. For more information about Wiley products, visit our web site at www.wiley.com.

Library of Congress Cataloging-in-Publication Data:

Keiser, Gerd.
    FTTX concepts and applications / by Gerd Keiser.
      p. cm. — (Wiley series in telecommunications and signal processing)
    "A Wiley-Interscience publication."
    Includes bibliographical references and index.
    ISBN-13: 978-0-471-70420-1 (alk. paper)
    ISBN-10: 0-471-70420-2 (alk. paper)
    ISBN-10: 0-471-76909-6 (e-book)
    1. Optical communications. I. Title. II. Series.

TK5103.59.K398   2006
621.382'7—dc22   2005050351

To Ching-yun and Nishla

Contents

Preface — xv

Chapter 1 Access Technologies — 1

1.1 General Network Concepts — 2
    1.1.1 Network Architecture Concepts — 2
    1.1.2 Types of Networks — 3
    1.1.3 Network Terminology — 4
    1.1.4 First-Mile Concept — 6
    1.1.5 Network Market Opportunities — 7
    1.1.6 Terminology for Premises — 8
1.2 Comparison of Access Technologies — 9
    1.2.1 Hybrid Fiber–Coax — 9
    1.2.2 Digital Subscriber Line — 10
    1.2.3 WiMAX — 12
1.3 Passive Optical Networks — 13
    1.3.1 Basic PON Architectures — 13
    1.3.2 What Is FTTx? — 14
1.4 Point-to-Point Links — 16
1.5 Summary — 17
    Further Reading — 17

Chapter 2 Optical Communications Essentials — 19

2.1 Definitions of Units and Terms — 19
    2.1.1 Metric Prefixes — 19
    2.1.2 Electromagnetic Spectral Bands — 20
    2.1.3 Optical Spectral Band — 21
    2.1.4 Digital Multiplexing Hierarchy — 22
    2.1.5 Decibel Units — 23
    2.1.6 Refractive Index — 26
2.2 Elements of an Optical Link — 26

2.3	Optical Fibers	28
	2.3.1 Fiber Structures	28
	2.3.2 Rays and Modes	31
2.4	Optical Fiber Attenuation	33
2.5	Fiber Information Capacity	35
	2.5.1 Modal Dispersion	35
	2.5.2 Chromatic Dispersion	36
	2.5.3 Polarization Mode Dispersion	37
2.6	Nonlinear Effects in Fibers	38
	2.6.1 Stimulated Brillouin Scattering	38
	2.6.2 Stimulated Raman Scattering	39
2.7	Optical Fiber Standards	40
2.8	Summary	41
	Problems	42
	Further Reading	44

Chapter 3 Wavelength-Division Multiplexing — 45

3.1	Operational Principles of WDM	46
	3.1.1 WDM Operating Regions	47
	3.1.2 Generic WDM Link	48
3.2	Standard WDM Spectral Grids	49
	3.2.1 Dense WDM	50
	3.2.2 Coarse WDM	50
	3.2.3 PON Spectral Regions	51
3.3	Optical Couplers	52
	3.3.1 Basic 2×2 Coupler	52
	3.3.2 Coupler Performance	53
	3.3.3 Tap Coupler	54
3.4	Bidirectional WDM Links	55
3.5	Summary	56
	Problems	57
	Further Reading	59

Chapter 4 PON Transceivers — 61

4.1	Optical Sources for PONs	62
	4.1.1 Source Characteristics	62
	4.1.2 DFB and FP Lasers	63
	4.1.3 Modulation Speed	64
	4.1.4 Optical Transmitter Packages	65
4.2	Optical Receivers	65
	4.2.1 Photodetector Types	66
	4.2.2 Quantum Efficiency	67

	4.2.3	Responsivity	67
	4.2.4	Speed of Detector Response	68
	4.2.5	Receiver Bandwidth	69
	4.2.6	Photodetector Noise	69
4.3	Receiver BER and OSNR	70
4.4	Burst-Mode Receiver Concept	71
4.5	Burst-Mode ONT Transmission	73
4.6	PON Transceiver Packages	74
4.7	Summary	75
	Problems	76
	Further Reading	77

Chapter 5 Passive Optical Components — 79

5.1	WDM Couplers for PONs	80
	5.1.1 Thin-Film Filters	81
	5.1.2 Transmission Diffraction Gratings	86
5.2	Optical Power Splitter	87
	5.2.1 Splitting Loss	88
	5.2.2 Optical Splitter Structure	88
5.3	Optical Cables for PONs	90
	5.3.1 Cable Structures	90
	5.3.2 Fiber and Jacket Color Coding	92
5.4	Fiber Interconnections	93
	5.4.1 Optical Connectors	93
	5.4.2 Connector Losses	95
	5.4.3 Optical Splices	97
5.5	Summary	98
	Problems	99
	Further Reading	100

Chapter 6 Passive Optical Networks — 101

6.1	Fundamental PON Architecture	102
6.2	Active PON Modules	104
	6.2.1 Optical Line Terminal	104
	6.2.2 Optical Network Terminal	105
	6.2.3 Optical Network Unit	105
6.3	Traffic Flows	106
6.4	Passive Component Applications	108
	6.4.1 Optical Cables for PONs	108
	6.4.2 Optical Power Splitters	108

	6.4.3	Splitter Enclosures	110
	6.4.4	Wavelength Couplers	110
6.5	PON Alternatives		111
	6.5.1	BPON Basics	112
	6.5.2	EPON and EFM	112
	6.5.3	GPON Basics	113
6.6	Optics Path Attenuation Ranges		113
6.7	Standards Development		114
	6.7.1	ITU-T	114
	6.7.2	FSAN	114
	6.7.3	IEEE	115
6.8	Summary		115
	Problems		116
	Further Reading		117

Chapter 7 BPON Characteristics — 119

7.1	BPON Architecture		119
	7.1.1	Traffic Flow Schemes	120
	7.1.2	OLT Capabilities	121
7.2	ATM Basics		122
	7.2.1	Use of ATM Cells	122
	7.2.2	ATM Service Categories	123
	7.2.3	Service Level Agreements	126
7.3	BPON Operational Characteristics		126
	7.3.1	Voice and Data Traffic Flows	127
	7.3.2	Protection of Grants	128
	7.3.3	Video Traffic	129
7.4	Traffic Control		129
	7.4.1	Fixed Bandwidth Allocation	130
	7.4.2	Dynamic Bandwidth Allocation	130
7.5	Standards Details		132
	7.5.1	Recommendation G.983.1	132
	7.5.2	Recommendation G.983.2	132
	7.5.3	Recommendation G.983.3	133
	7.5.4	Recommendation G.983.4	133
	7.5.5	Recommendation G.983.5	133
	7.5.6	Recommendation G.983.6	133
	7.5.7	Recommendation G.983.7	134
	7.5.8	Recommendation G.983.8	134
7.6	Summary		134
	Problems		135
	Further Reading		137

Chapter 8 Ethernet in the First Mile — 139

8.1 EFM Options — 140
8.2 EPON Architecture — 141
    8.2.1 OLT and ONT/ONU Functions — 142
    8.2.2 EPON Traffic Flows — 142
    8.2.3 Power Levels Received — 145
8.3 MPCP Functions — 145
    8.3.1 Discovery Process — 145
    8.3.2 Bandwidth Assignment — 146
    8.3.3 Transmission Timing — 147
8.4 Point-to-Point Ethernet — 148
    8.4.1 P2P Ethernet Over Fiber — 148
    8.4.2 P2P Ethernet Over Copper — 149
8.5 Main EPON and P2P EFM Standards — 149
8.6 Summary — 150
    Problems — 150
    Further Reading — 153

Chapter 9 GPON Characteristics — 155

9.1 GPON Architecture — 155
    9.1.1 GSR Specification — 156
    9.1.2 GPON Protection Switching — 157
    9.1.3 Information Security in a GPON — 158
9.2 GPON Recommendation G.984.2 — 159
    9.2.1 Optical Performances — 159
    9.2.2 Timing and Optical Power Control — 160
    9.2.3 Forward Error Correction — 161
9.3 GPON Transmission Convergence Layer — 162
    9.3.1 Downstream GPON Frame Format — 162
    9.3.2 Upstream GPON Frame Format — 164
    9.3.3 GEM Segment — 165
9.4 ONT Management and Control — 166
9.5 Summary — 166
    Problems — 167
    Further Reading — 169

Chapter 10 FTTP Concepts and Applications — 171

10.1 Implementation Scenarios — 171
    10.1.1 Application Alternatives — 171
    10.1.2 Installation Types — 173

xii CONTENTS

10.2	Network Architectures	175
	10.2.1 Optical Splitter Locations	175
	10.2.2 Network Design Variations	177
10.3	Local Powering Options	179
	10.3.1 Indoor Power Supply	179
	10.3.2 Outdoor Power Supply	180
	10.3.3 Network Powering	181
10.4	Service Applications	181
	10.4.1 Bandwidth Requirements	182
	10.4.2 Video Service Issues	182
10.5	Expanded WDM PON	184
10.6	Summary	185
	Problems	186
	Further Reading	187

Chapter 11 FTTP Network Design — 189

11.1	Design Criteria	189
	11.1.1 System Requirements	190
	11.1.2 System Margin	191
	11.1.3 Power Penalties	191
11.2	Link Power Budget	193
	11.2.1 Power-Budgeting Process	194
	11.2.2 FTTP 1310-nm Power Budget	196
	11.2.3 FTTP 1490-nm Power Budget	198
11.3	Photonic Design Automation Tools	199
	11.3.1 Modeling Tool Characteristics	199
	11.3.2 FTTP Network Modeling Tool	200
11.4	Link Capacity Estimates	200
	11.4.1 Basic Formulation	200
	11.4.2 Basic Rise Times	201
	11.4.3 FTTP Link Rise Time	202
11.5	Network Protection Schemes	203
11.6	Summary	204
	Problems	205
	Further Reading	207

Chapter 12 FTTP Network Implementations — 209

12.1	Central Office Configuration	209
	12.1.1 Service Inputs to the FTTP Network	210
	12.1.2 Cable Layout and Interfaces	211

CONTENTS **xiii**

		12.1.3	WDM Coupler Placement	212
		12.1.4	Patch Cords and Intrafacility Cables	214
	12.2	Feeder Cables		215
		12.2.1	Feeder Cable Structures	215
		12.2.2	OSP Distribution Cabinet	216
	12.3	Distribution Section		217
	12.4	Installation of PON Cables		219
		12.4.1	Direct-Burial Installations	220
		12.4.2	Horizontal Drilling	223
		12.4.3	Pulling Cable into Ducts	224
		12.4.4	Cable Jetting Installation	225
		12.4.5	Aerial Installation	228
		12.4.6	Cable Warning and Identification Markers	228
	12.5	Summary		230
		Problems		231
		Further Reading		232

Chapter 13 Network Installation Testing   233

13.1	International Measurement Standards		235
13.2	Basic Test Instruments		236
13.3	Optical Power Measurements		237
	13.3.1	Definition of Optical Power	237
	13.3.2	Optical Power Meter	238
	13.3.3	Power Meter Applications	239
13.4	Optical Time-Domain Reflectometer		240
	13.4.1	OTDR Trace	240
	13.4.2	OTDR Dead Zone	242
	13.4.3	Fiber Fault Location	243
13.5	Optical Return Loss		243
13.6	Visual Fault Locator		244
13.7	Optical-Loss Test Set		245
13.8	Multifunction Test Instrument		245
13.9	Device Conformance Testing		246
13.10	FTTP Network Testing		247
	13.10.1	Checking Individual Link Losses	248
	13.10.2	Optical-Loss Budget Check	249
	13.10.3	End-to-End Link Characterization	249
	13.10.4	ORL Measurements	251
	13.10.5	OLT and Video Output Checks	251
	13.10.6	ONT Output Check	252

13.11	FTTP Network Troubleshooting		252
	13.11.1 Resolutions of Network Problems		253
	13.11.2 Troubleshooting Guidelines		255
13.12	Summary		255
	Problems		256
	References and Further Reading		258

Chapter 14 Network Management Functions   259

14.1	Basic Network Management	260
14.2	Management Functions	261
	14.2.1 Performance Management	262
	14.2.2 Configuration Management	262
	14.2.3 Accounting Management	263
	14.2.4 Fault Management	263
	14.2.5 Security Management	264
14.3	OAM&P in FTTP Networks	265
14.4	Summary	266
	Problems	267
	Further Reading	268

Appendix A	**Units, Physical Constants, and Conversion Factors**	**269**
Appendix B	**Acronyms**	**271**
Appendix C	**Video Transmission**	**277**
Appendix D	**Communication Signals**	**279**
Appendix E	**Telcordia Generic Requirements for PON Applications**	**283**

Index   **285**

Preface

Prior to widespread use of the Internet, telecommunication service customers used only standard telephones, fax machines, or dial-up modems to communicate on a worldwide basis. To connect to the outside world, these applications typically used the public switched telephone network (PSTN), which consists of twisted-pair copper wire links that run from customer premises to local telecommunication switching (distribution) centers. Except for occasional holidays when large usage peaks can occur, this PSTN traffic flow followed regular and predictable patterns with limited connection times. In this scenario, telephone companies concentrated on building high-capacity networks up to a local distribution point. Beyond that point, connections to the customers' premises were by lower-speed copper wires.

This situation changed with the advent of powerful personal computers, which created a demand for new bandwidth-hungry applications and services, each of which could consume several megabits per second. In addition to large usage demands, the corresponding traffic patterns tend to be bursty and unpredictable in nature, and the connection times became much longer than in a PSTN. This required a new look at the capabilities of the access network, which encompasses the connections that extend from a local switching facility to individual businesses, organizations, and homes.

This book presents fundamental passive optical network (PON) concepts needed to understand, design, and build high-capacity links for access networks. The sequence of topics takes the reader systematically from the underlying principles and components of optical fiber communication links, through descriptions of fundamental PON types and operations, to their application to fiber-to-the-premises (FTTP) networks, and finally, to essential measurement and test procedures required during network installation and maintenance. Key features of the book for accomplishing this are:

- A comparison of several competing technologies used for high-speed access networks.
- A detailed overview of the types and behavior of optical fibers, cables, couplers, connectors, and transceivers needed to implement FTTP networks.
- A discussion of the fundamental constituents and operations of a PON, such as triple-play services over a single fiber, the use of optical power splitters, and the functions of PON transmission equipment.

- Detailed descriptions of the architectural differences of three basic PON types, the standards governing them, and the rationales for their development.
- Illustrations of FTTP implementation concepts and criteria in various categories of neighborhoods, optical power-budget analyses for various network designs, and options for supplying electrical power to optical network units and optical network terminations.
- Discussions of FTTP cable plant implementations, including cabling interfaces and enclosures in the central office, types of optical cables and distribution cabinets used in the outside plant, and fiber cable installation procedures.
- Descriptions of measurement techniques for verifying that the network has been configured properly and that its constituent components are functioning correctly.
- An overview of the issues related to the operation, administration, maintenance, and provisioning functions of a PON.
- Web access through the book website to an interactive FTTP network simulation tool for educational purposes (see ftp://ftp.wiley.com/public/sci_tech_med/fttx_concepts/).

This book provides the basic material for an introductory senior-level course on the principles and applications of passive optical network technology. It also will serve well as a working reference for practicing engineers and managers dealing with FTTP network design, implementation, installation, testing, and maintenance. Others who can benefit from the material in the book include service provider management and engineering personnel, optoelectronic device and module manufacturers who want a better understanding of how their devices fit into FTTP networks, investment personnel who need to know the technical and implementation issues related to FTTP networks, and technical sales and marketing people.

Since modern communication networks make extensive use of acronyms, a list of the most common ones is given at the end of the book. A collection of 126 homework problems is included to help test the reader's comprehension of the material covered and to extend and elucidate the text. A number of the homework problems can be expanded to serve as student projects.

Simulation tools are being used extensively for modeling the behavior of local, metro, and long-haul networks. These tools also can be useful in the physical-layer design of FTTP networks. The book Web site gives an example application note of such a tool and related applications from RSoft Design Group. In addition, there is a link to the company Web site (www.rsoftdesign.com) that will allow readers access to a downloadable interactive model for an example FTTP network.

In preparing this book, many thanks go to the manuscript reviewers, whose comments enhanced and clarified the content and organization of the material. An initial trial run of the material in this book was presented at a 2004 summer course at the National Cheng Kung University (NCKU) in Tainan, Taiwan. Special thanks go to San-Liang Lee of the National University of Science and Technology and Hen-Wai Tsao of the National Taiwan University for inviting me to give this course, to Jen-Fa

Huang of NCKU and his wife Dailie Cheng for their gracious hospitality during the course, and to Hui-Ru Lin and Wan-Ting Liao for their extensive help with the logistics of this course.

In addition, I am greatly indebted to the numerous people with whom I had many discussions and who supplied me with photos and other material. Among them are Greg Slabodkin of ADC; Larry Woody of Charles Industries; Paul Fitzgerald and Paul Fowler of Circadiant Systems; André Girard of EXFO; Jason Greene of Furukawa America; Rosmin Robertson of Kingfisher; Joy Jiang of Lightwaves2020; Ley Mee Hii of LuminentOIC; Jeff Sitlinger of OFS; Brent Whitlock of RSoft Design Group; Frank Jaffer of SENKO Advanced Components; Robert Orr of Sherman & Reilly; Evie Bennett of William Frick and Company; and Jessica Held of Vermeer. This book especially benefited from the expert guidance of George Telecki, Rachel Witmer, and Angioline Loredo at Wiley. It was truly a pleasure to work with them. As a final personal note, I am grateful to my wife, Ching-yun, and my daughter, Nishla, for their patience, help, and encouragement during the time I devoted to writing this book.

GERD KEISER

Newton Center, MA

1

ACCESS TECHNOLOGIES

The emergence and dramatic surge in use of the World Wide Web (commonly known as the *Web*) in the 1990s greatly changed the fundamental nature of network design and use. Before that time the main focus of network providers was to increase the utilization of long-haul telephone links by multiplexing many low-speed users onto high-speed large-capacity fiber optic links. Telephone companies have spent large sums of money to build out such networks up to a local distribution point. Beyond that point the connections to customers' premises were lower-speed copper wires. The cable system that interconnects the user sites and the huge network beyond a local telecommunication distribution point (such as a telephone switching station) is called the *access network*. The cost of the large-capacity long-haul networks, which requires the installation of expensive sophisticated switching equipment, is spread out over a large number of users, so the relative cost per service subscriber is low.

Initially, end customers used only telephones, fax machines, or dial-up modems, for which inexpensive copper-based twisted-wire links from their premises to the outside distribution networks were sufficient. This scenario changed with the advent of powerful personal computers, which created a demand for new bandwidth-hungry applications and services, each running at rates of several megabits per second (Mbps). Among these high-rate applications are video on demand, streaming media, virtual private circuits, high-resolution image transfers, and online entertainment in addition to traditional phone, data, and fax services. These new demands drove the need to replace low-capacity copper access links with higher-capacity connections. Several competing broadband wireline and wireless telecommunication technologies are addressing the demand for faster access links. They include hybrid fiber–coax networks, digital subscriber line technology, broadband wireless links, free-space optical links, and passive optical network schemes.

In this chapter we give an overview of access network concepts, terms, and interfaces in Section 1.1. Next, in Section 1.2 we present a comparison between the

---

FTTX Concepts and Applications, by Gerd Keiser
Copyright © 2006 John Wiley & Sons, Inc.

access network methodologies that compete with passive optical network (PON) schemes. Section 1.3 gives a snapshot of what a PON is, and in Section 1.4 we introduce the enabling technologies and the Ethernet-in-the-first-mile concept. In subsequent chapters we describe the fundamental concepts and applications of various PON types that are being used and considered for access networks.

1.1 GENERAL NETWORK CONCEPTS

To demonstrate how an access network fits into the overall networking hierarchy, in this section we illustrate different types of networks, note who owns and operates them, and define some network terminology.

1.1.1 Network Architecture Concepts

The term *architecture* is used widely when discussing the design and implementation of a telecommunication network. An architecture is basically a set of rules and conventions by which something is built. Thus, in telecommunication systems the *network architecture* describes the general physical arrangement and operational characteristics of communicating equipment together with a common set of communication protocols. A *protocol* is a set of rules and conventions that governs the generation, formatting, control, exchange, and interpretation of information that is transmitted through a telecommunication network or that is stored in a database.

A traditional approach to setting up a protocol is to subdivide it into a number of individual pieces of manageable and comprehensible size. The result is a layered structure of services, which is referred to as a *protocol stack*. In this scheme each layer is responsible for providing a set of functions or capabilities to the layer above it by using the functions or capabilities of the layer below. A user at the highest layer is offered all the capabilities of the lower levels for interacting with other users and peripheral equipment distributed on the network.

The seven-layer OSI (Open System Interconnect) reference model is an example of a structured approach. As shown in Figure 1.1, by convention these layers are viewed as a vertical sequence, with the numbering starting at the bottom layer. An increasingly larger number of functions are provided when moving up the protocol stack, and the level of functional abstraction increases. The lower layers govern the *communication facilities*, which deal with the physical connections, data-link control, and routing and relaying functions that support the actual transmission of data. The upper layers support *user applications* by structuring and organizing data for the needs of the user.

Note that there is nothing inherently unique about using seven layers or about the specific functionality in each layer. In actual applications some of the layers may be omitted and other layers can be subdivided into further sublayers. Thus, the layering mechanism should be viewed as a framework for discussions of implementation, not as an absolute authority. For PON discussions we are concerned primarily with the lower three levels.

GENERAL NETWORK CONCEPTS 3

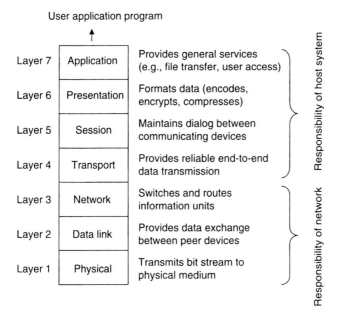

Figure 1.1. General structure and functions of the seven-layer OSI reference model.

1.1.2 Types of Networks

Figure 1.2 illustrates four generic categories of telecommunication networks.

1. A *wide area network* (WAN) spans a large geographical area. The links can range from connections between switching facilities in neighboring cities to long-haul terrestrial or undersea transmission lines running across a country or between countries. WANs are owned and operated by either private enterprises or by telecommunication service providers. Invariably, many companies own the various segments of a WAN link from one distant point to another.

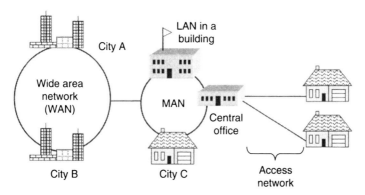

Figure 1.2. Broad categories of networks ranging from LANs to WANs.

2. A *metropolitan area network* (MAN) spans a smaller area than that spanned by a WAN. This could range from interconnections between buildings covering several blocks within a city or could encompass an entire city and the metropolitan area surrounding it. There is also some means of interconnecting the MAN resources with communication entities located in both WANs and local area networks. MANs are owned and operated by many organizations. They commonly are referred to as *metro networks*.

3. A *local area network* (LAN) interconnects users in a localized area such as a room, a department, a building, an office or factory complex, or a campus. Here the word *campus* refers to any group of buildings that are within reasonable walking distance of each other. For example, it could be the collocated buildings of a corporation, a large medical facility, or a university complex. LANs usually are owned, used, and operated privately by a single organization.

4. An *access network* encompasses connections that extend from a central communication switching facility (called the *central office*; see Section 1.1.3) to individual businesses, organizations, and homes. One of its functions is to collect and concentrate the information flows from customer locations and then send this aggregated traffic to the central office. In the other transmission direction, access networks allow carriers to provide voice, data, video, and other services to subscribers. Such access networks typically are owned by service providers. Access networks thus enable users in homes, businesses, and other enterprises to have connections to global information transport networks. The main drive behind new access technology developments is to transition from slow traditional telephone lines to large-capacity links to allow users to have high-speed connections to outside networks.

When a network is owned and deployed by a private enterprise, it is referred to as an *enterprise network*. For a monthly fee, the networks owned by the telecommunication carriers provide services such as leased lines or real-time telephone connections to other users and to enterprises at large. Such networks are referred to as *public networks*.

1.1.3 Network Terminology

To get a better understanding of optical networks, we need to define some terms used in a public network, such as that shown in Figure 1.3.

- *Central office.* A communication switching facility in a public network is called a *central office* (CO) or a *point of presence* (POP). A CO may be located on a metro ring or on a link running to the core metro network. It houses a series of large complex switches that establish temporary connections for the duration of a requested connection time between subscriber lines which terminate at the switch. The CO can serve on the order of thousands of subscribers.
- *Long-haul network.* A *long-haul network* interconnects various cities or geographical regions and spans hundreds to thousands of kilometers between central offices.

GENERAL NETWORK CONCEPTS   5

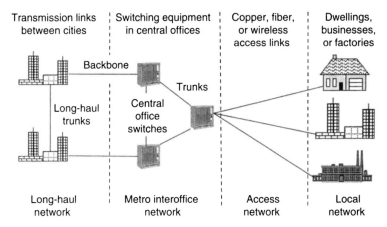

Figure 1.3. Definitions of some terms used in describing a public network.

- *Backbone network*. The term *backbone* describes a high-capacity network that connects multiple LAN, MAN, or WAN segments. Thus, a backbone handles internetwork traffic, that is, traffic that originates in one segment and is transmitted to another segment.
- *Metro interoffice network*. A metropolitan (typically abbreviated *metro*) network connects groups of central offices within a city or geographical region. The distances between central offices for this type of network range from a few to several tens of kilometers.

The access network is the topic of interest for PON applications. Note that to distinguish the traffic-flow directions, the term *downstream* refers to transmissions from the central office to users, and the term *upstream* is the transmission path from users to the central office. The following are some terms used to describe equipment locations and cable types within this network segment, as illustrated in Figure 1.4:

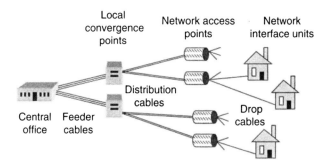

Figure 1.4. Some terms used to describe equipment locations and cable types within a PON.

- *Local convergence point.* The term *local convergence point* (LCP) refers to access telecommunication assets located in a neighborhood or business park. These assets serve many subscribers and usually consist of an indoor or outdoor cabinet or enclosure that houses optical devices or electronic equipment. The function of the equipment is to distribute signals downstream to the customer or to concentrate individual customer information flows and transmit the aggregated traffic upstream to the central office.
- *Feeder cables.* The links connecting the CO and the LCP are known as *feeder cables.*
- *Network access point.* The *network access point* (NAP) or *access terminal* is located close to subscribers in a neighborhood, business park, or other campus. This serves as a concentration or distribution point for breaking out short links to individual customer's premises.
- *Distribution and drop cables.* The *distribution cables* originate at the LCP and connect to many network access points. At the network access point, the wires or fibers in the distribution cables are joined to individual *drop cables* that connect to customers' premises.
- *Network interface unit.* A *network interface unit* (NIU) located at the customer site enables the user to connect to the network. Typically, the NIU will contain electronics that serve as the interface mechanism between the transmission medium of the access network and the particular wiring inside the subscriber premises, for example, optical fiber, twisted-pair copper wires, or coaxial cable. As shown in later chapters, this unit may have various names, depending on the technology used for the access network.

1.1.4 First-Mile Concept

Figure 1.5 shows a simplified access network that allows the customer to subscribe to phone, video, and Internet services. In a PON a customer's premises are connected by means of a passive optical link to the central office, which interfaces to telecommunication services such as the public switched telephone network (PSTN), Internet service providers (ISPs), video-on-demand providers, or a storage area network (SAN).

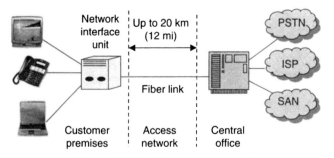

Figure 1.5. Simplified access network.

The terms *last mile* or *first mile* are used interchangeably to refer to the span in an access network between a business or a residence and the local central office. Note that the term *mile* is used merely to denote a relatively short distance and does not mean exactly 1 mile. As described in Chapter 6, this distance can be up to 20 km (12 miles). Whether this distance is called the *first mile* or the *last mile* depends on the direction from which one views it. The phone companies call it the *last mile*, since it represents the final connection from the CO to a customer. The vendors of Ethernet equipment prefer to call it the *first mile*, since it is the initial connection from a customer's equipment to the network.

The *first-mile issue* or *last-mile issue* is related to performance limitations of the traditional inexpensive low-capacity media used for this span. For large-capacity wide area and metro networks, the cost of installing and operating extensive and expensive cabling and switching equipment can be spread out over a large number of users, so that the relative cost per service subscriber is low. However, running a high-speed link from a local convergence point to the premises is expensive in relation to an individual user, since the implementation cost cannot be shared among many customers.

1.1.5 Network Market Opportunities

Different market opportunities exist for access networks, depending on whether the area being served is a new or an existing neighborhood. As Figure 1.6 illustrates, the three key market segments are known as greenfield, overbuild, and refurbishment (or access rehabilitation) applications. The term *greenfield* applies to new housing or business park developments. Installing new fiber-based access networks in such developments is very attractive to large service providers, since optical fiber cables or ducts can be placed easily and inexpensively into existing construction trenches throughout the new neighborhood. With the cable or ducts in place, fibers can be run to each new building. Although not every homeowner or business may necessarily subscribe to such a network, the chances are that most will. As an additional cost

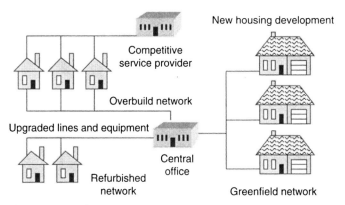

Figure 1.6. The three key PON market segments are known as greenfield, overbuild, and refurbishment applications.

8 ACCESS TECHNOLOGIES

saving to the service provider, since having a high-speed access to each home is a competitive selling point, the developer may decide to absorb some of the cost of installing the access network.

The term *overbuild* refers to an additional service provider installing a new broadband network capability in a community that is already being served by other competing organizations, such as one or more cable operators or digital subscriber line providers. Thus, it sometimes is referred to as a *competitive overbuild*. Typically, the implementing organization will use an overbuild network to deliver more capacity and better service than any of the incumbent service providers can offer.

Refurbishment (sometimes called *rehabilitation*) of a network refers to replacing low-capacity copper wires with larger-capacity optical fibers in an established neighborhood. Such an upgrade and expansion of the access network not only creates new revenue-generating opportunities because of the higher-capacity lines, but also greatly reduces maintenance time and cost that would have been associated with an older network that may be prone to failures or degradations.

1.1.6 Terminology for Premises

The end user in a first-mile network can be either in a single-family dwelling, a single-business unit, or a building that houses multiple apartments, businesses, or other organizations. Figure 1.7 shows some of these premises types. The following acronyms are used to refer to variations on premises that contain multiple groups of users:

- A *multiple dwelling unit* (MDU) refers to apartment complexes, condominiums, or dormitories.
- A *multiple tenant unit* (MTU) designates an office building, an office campus, or an industrial campus with different business tenants.

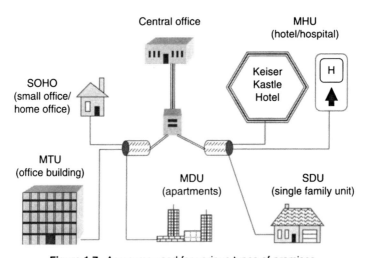

Figure 1.7. Acronyms used for various types of premises.

- A *multiple hospitality unit* (MHU) refers to premises such as hotels, hospitals, airports, or convention centers.
- A *single dwelling unit* (SDU) is a premises occupied by a single family.
- A *single family unit* (SFU) is an alternative designation of an SDU.
- A *small office/home office* is referred to by the acronym SOHO.

1.2 COMPARISON OF ACCESS TECHNOLOGIES

Several different wireline and wireless technologies have been used for high-capacity links in the access network. The ones addressed in this section are hybrid fiber–coax schemes, digital subscriber lines, and broadband wireless links. These technologies are competitors of passive optical networks (PONs) and high-speed point-to-point (popularly called P2P) access links, which are the topic of this book. For comparison purposes, in Section 1.3 we give a brief overview of PONs and introduce the use of P2P access links in Section 1.4. In Chapter 6 we give more information on PONs and the general concept of P2P links. In Chapters 7 through 9 we present more details on the various types of PON architectures.

1.2.1 Hybrid Fiber–Coax

The use of a combination of optical fibers and coaxial cable is known as a *hybrid fiber–coaxial* (HFC) system. This scheme has been implemented widely by cable TV (CATV) companies. HFC networks deploy fiber optic cable and coaxial cable in different portions of an access network to carry broadband content, such as video, data, voice, and Internet files. As shown in Figure 1.8, a local CATV company runs a fiber optic cable from an information distribution center (called the *cable headend*) to serving nodes, which are located close to business and residential users. At this point the optical signals are converted to electrical signals, which then travel over coaxial cables to individual businesses and homes.

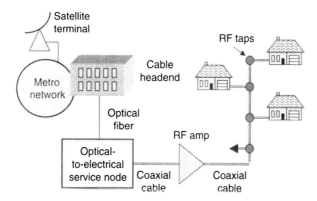

Figure 1.8. Hybrid fiber–coax network.

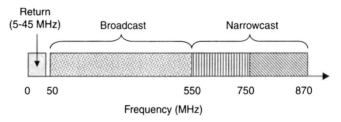

Figure 1.9. Frequency spectrum used by an HFC network.

In HFC systems the terms *forward* and *reverse* (or *return*) are used to refer to the downstream and upstream directions, respectively. The coaxial wires used in HFC networks allow the transport of broadband information over distances of several tens of kilometers. The cable attenuation over such distances is compensated by the periodic use of inexpensive RF (radio-frequency) amplifiers. To attach a new user to the HFC system, the service provider uses a simple coaxial T-connector to tap into the main line in order to run a drop cable to the premises. To access the services on the network, the customer uses a relatively inexpensive modem that attaches to the drop cable and separates the TV and data signals internally.

Figure 1.9 shows the frequency spectrum used by an HFC network. The 40-MHz-wide band ranging from 5 to 45 MHz is used to transmit information from the user back to the headend. The 500-MHz span ranging over the frequencies from 50 to 550 MHz is used for analog broadcast, such as analog video distribution or FM sound. The narrowcast band starting at 550 MHz is either 200 or 320 MHz wide, depending on the system. This band is used for services such as video on demand, high-definition TV (HDTV), and multimedia applications.

1.2.2 Digital Subscriber Line

Standard telephone installations consist of pairs of twisted copper wires that run from a central office to homes and businesses. Although such wires are capable of sending signals in a 1-MHz bandwidth, traditionally the telephone service network used only a 4-kHz bandwidth. The sudden large Internet-generated demand for data services in addition to telephone capabilities created the concept of *digital subscriber line* (DSL) technology. Since the copper wires were already in place almost everywhere, service providers decided to send data in the unused 4-kHz to 1-MHz bandwidth of these wires.

A number of different variations of DSL have appeared. Table 1.1 lists some of the main ones and gives their characteristics. The principal variation is *asymmetric DSL* (ADSL). The International Telecommunications Union (ITU) has approved a set of standards for ADSL, which are numbered from G.992.1 through G.992.5. Figure 1.10 shows a basic ADSL architecture. The word *asymmetric* means that the ADSL scheme transmits data at two different rates. The downstream information flowing from the central office to the customer is sent at a higher rate than the upstream traffic sent from the customer to the network. The reason for this asymmetry is that

COMPARISON OF ACCESS TECHNOLOGIES 11

TABLE 1.1 Variations of Digital Subscriber Line Technologies

DSL Type	Downstream Rate (Mbps)	Upstream Rate (Mbps)	Access Range (ft/km)	Characteristics
ADSL	1.7 7.1	0.176 0.640	18,000/5.5 12,000/3.6	The long- and short-range options can cover suburban areas of 38 or 95 km^2, respectively. This allows services to 20,000 or 100,000 customers.
HDSL	1.544 2.048	1.544 2.048	18,000/5.5 12,000/3.6	The two rates of this symmetric service correspond to the T1 and E1 telephone rates used in the SONET and SDH protocol hierarchies, respectively (see Section 2.1)
SHDSL	0.192 2.3 4.62	0.192 2.3 4.62	40,000/12.2 6500/2.0 6500/2.0	There are three classes of symmetric service rates for distances up to 12.2 km.
VDSL	13 26	0.640 3.2	4500/1.37 300/0.91	This method provides a very high data rate but has a limited distance-sensitive access range.

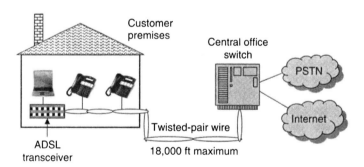

Figure 1.10. Basic ADSL architecture.

typically, customers use the upstream link mainly for sending information requests or relatively short messages. In the other direction, the traffic flowing to the customer can consist of items requiring a much higher bandwidth, such as large data files, video clips, or high-resolution product information from the Internet. Upstream rates range from 16 to 640 kbps, and downstream information flows at speeds of 1.5 to 8 Mbps.

In the frequency spectrum used by ADSL, the 0- to 4-kHz band is used for voice traffic, and the frequency band from 10 kHz to 1 MHz is used for data traffic. These separate bandwidth channels allow customers to send voice and data traffic simultaneously. As described in G.992.1, most ADSL equipment uses a signaling system called *discrete multitone* (DMT). The DMT scheme divides the data band into 247 separate 4-kHz channels. The ADSL performance control monitors each channel, and if the quality degrades the system, switches the signal to a different channel.

Although ADSL links are inexpensive to implement since they use existing telephone networks, the transmission characteristics of the phone lines impose several limitations on ADSL. First, the copper wire links were designed and optimized for carrying voice traffic in the very narrow frequency band ranging from 0 to 4 kHz. To achieve this, the telephone lines use special filters to condition the voice signals and to eliminate frequencies greater than 4 kHz. In addition, the telephone companies periodically use small amplifiers called *loading coils* along the line to boost the strength of voice signals. Since these loading coils are incompatible with ADSL, a customer cannot receive ADSL service if such an amplifying coil is located in the line between the central office and the customer.

These factors also cause ADSL to be *distance sensitive*. The distance limit for receiving ADSL service is 18,000 ft (5460 m) or 3.4 miles. In addition, the connection speed to the customer decreases as the distance to the central office increases. For users located less than 6000 ft (1820 m) away, the connection speeds can be up to 8 Mbps downstream and 640 kbps upstream. However, customers who are farther away typically can receive only at rates of 1.5 Mbps and send at speeds varying between 64 and 640 kbps.

1.2.3 WiMAX

WiMax (Worldwide Interoperability for Microwave Access) is the popular term for the IEEE Standard 802.16, which is called *Air Interface for Fixed Broadband Wireless Access Systems*. Ideally, a WiMax system can provide connectivity at speeds of up to 70 Mbps over distances as far as 30 miles. The initial implementation is being done by means of a fixed network of base stations, which is described in IEEE Standard 802.16-2004. WiMax also will support wireless transmissions directly to mobile end users, which is the topic of IEEE Standard 802.16e.

Figure 1.11 illustrates a typical fixed WiMAX network configuration. Here a backbone of base stations is connected to a public network by means of point-to-point links. Each base station can support hundreds of fixed WiMAX subscriber stations. These can be located in residences, businesses, or IEEE 802.11-based Wi-Fi (*wireless fidelity*) *hot spots*, which are short-distance public wireless access points to the Internet in places such as airports, hotels, and fast-food facilities. WiMAX for fixed access operates in a frequency band running from 2 to 11 GHz. Many of these frequencies do not require a license, such as the 5.8-GHz license-exempt band. Licensed bands include the 2.5- and 3.5-GHz regions. The antenna for the subscriber equipment is mounted on a roof or mast, similar to the installation of a small satellite dish. The IEEE 802.16-2004 standard also has the option of installing the antenna indoors, although transmissions from such a location may not be as reliable as outdoor installations.

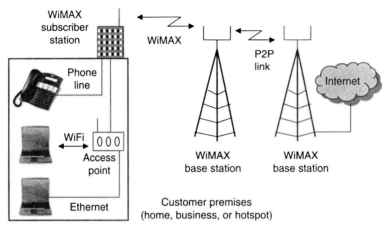

Figure 1.11. Typical fixed WiMAX network configuration.

The 802.16-2004 Standard uses a *grant-request access protocol*, and the WiMAX base station coordinates all communications. This guarantees that there are no data collisions between users. Thus, the available bandwidth is used more efficiently than in a contention-based network such as WiFi. The base stations use the media access control procedure defined in the standard to allocate upstream and downstream bandwidth to subscribers according to their needs on a real-time basis.

1.3 PASSIVE OPTICAL NETWORKS

Given that network and service providers are seeking to reduce their operational costs, the concept of using a *passive optical network* (PON) is an attractive option. In a PON there are no active components between the central office and the customer's premises. Instead, only passive optical components are placed in the network transmission path to guide the traffic signals contained within specific optical wavelengths to the user endpoints and back to the central office. Replacing active devices with passive components provides a cost savings to the service provider by eliminating the need to power and manage active components in the cable system of the access network. In addition, since the passive devices have no electrical power or signal-processing requirements, they have virtually an unlimited *mean time between failures* (MTBF). This obviously lowers the overall maintenance costs significantly for the service provider.

1.3.1 Basic PON Architectures

Figure 1.12 illustrates the architecture of a typical PON in which a fiber optic network connects switching equipment in a central office with a number of service subscribers. Examples of equipment in the central office include public telephone switches, video-on-demand servers, Internet protocol (IP) routers, Ethernet switches, and asynchronous transfer mode (ATM) switches. In the central office, data and

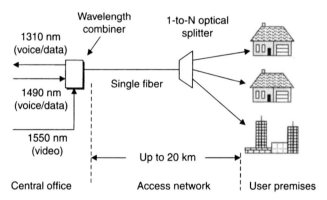

Figure 1.12. Architecture of a typical passive optical network.

digitized voice are combined and sent downstream to customers over an optical link using a 1490-nm wavelength. The upstream (customer to central office) return path for the data and voice uses a 1310-nm wavelength. Video services are sent downstream with a 1550-nm wavelength. There is no video service in the upstream direction. The transmission equipment in the network consists of an *optical line terminal* (OLT) situated at the central office and an *optical network terminal* (ONT) at each customer's premises.

Starting at the central office, one single-mode optical fiber strand runs to a passive *optical power splitter* near a housing complex, an office park, or some other campus environment. At this point the splitting device simply divides the optical power into N separate paths to the subscribers. If the splitter is designed to divide the incident optical power evenly, and if P is the optical power entering the splitter, the power level going to each subscriber is P/N. Designs of power dividers with other splitting ratios are also possible, depending on the application. The number of splitting paths can vary from 2 to 64, but typically, is 8, 16, or 32. From the optical splitter, individual single-mode fibers then run to each building or to the serving equipment. The optical fiber transmission span from the central office to the user can be up to 20 km. Thus, active devices exist only in the central office and at the user end.

There are several alternative PON implementation schemes. The three primary ones are broadband PON (BPON), Ethernet PON (EPON), and gigabit PON (GPON). In addition, occasionally there will be reference to ATM PON (APON), a subset of the BPON category. The three PON types are described in Chapters 7, 8, and 9, respectively.

1.3.2 What Is FTTx?

The application of PON technology for providing broadband connectivity in the access network to homes, multiple-occupancy units, and small businesses commonly is called *fiber-to-the-x*. This application is given the designation FTTx. Here x is a letter indicating how close the fiber endpoint comes to the actual user. Figure 1.13 illustrates

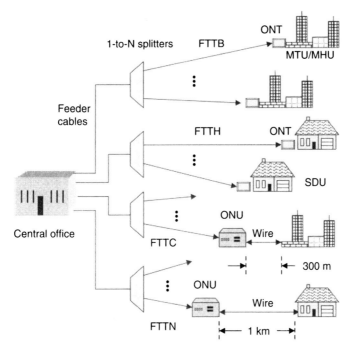

Figure 1.13. Some FTTx scenarios.

some of these scenarios. Among the acronyms used in the technical and commercial literature are the following:

- FTTB, *fiber-to-the-business*, refers to the deployment of optical fiber from a central office switch directly into an enterprise.
- FTTC, *fiber-to-the-curb*, describes running optical fiber cables from central office equipment to a communication switch located within 1000 ft (about 300 m) of a home or enterprise. Coaxial cable, twisted-pair copper wires (e.g., for DSL), or some other transmission medium is used to connect the curbside equipment to customers in a building.
- FTTH, *fiber-to-the-home*, refers to the deployment of optical fiber from a central office switch directly into a home. The difference between FTTB and FTTH is that typically, businesses demand larger bandwidths over a greater part of the day than do home users. As a result, a network service provider can collect more revenues from FTTB networks and thus recover the installation costs sooner than for FTTH networks.
- FTTN, *fiber-to-the-neighborhood*, refers to a PON architecture in which optical fiber cables run to within 3000 ft (about 1 km) of homes and businesses being served by the network.
- FTTO, *fiber-to-the-office*, is analogous to FTTB in that an optical path is provided all the way to the premises of a business customer.

- FTTP, *fiber-to-the-premises*, has become the prevailing term that encompasses the various FTTx concepts. Thus FTTP architectures include FTTB and FTTH implementations. An FTTP network can use BPON, EPON, or GPON technology.
- FTTU, *fiber-to-the-user*, is the term used by Alcatel to describe their products for FTTB and FTTH applications.

Of all these acronyms, the prevailing terms are FTTH and FTTP. As described in Section 10.3, an implementation complexity of FTTB and FTTH, as opposed to FTTC or FTTN, is that electric power cannot be supplied to a premises along with the signal on the fiber optic lines. On the other hand, FTTC and FTTN may require more maintenance time and costs, since they have active equipment in outside plant locations.

1.4 POINT-TO-POINT LINKS

An architecture consisting of point-to-point Ethernet links is a competing access network technology. In Chapter 8 we give details on this access scheme.

As shown in Figure 1.14, one option for a point-to-point Ethernet optical access network is to have dedicated fibers running between a central office and Ethernet switches designated for individual subscribers. Here downstream voice, data, and video services are combined onto a 1550-nm wavelength, and upstream services back to the central office are sent over the same fiber by means of a 1310-nm wavelength. The links can operate at 10 Gbps over distances up to 10 km. Such a scenario requires a large number of optical fiber lines, with each line having its own optical transceivers. For example, suppose that the network serves 16 subscribers. If the optical link running to an individual subscriber is bidirectional, 16 fibers are required. In case the links are unidirectional, a total of 32 optical fibers are needed. Since each subscriber link needs transmitters and receivers at each end, the system needs a total of 32 optical transceivers. Therefore, this type of network is useful only if each subscriber requires close to the full capacity offered by a gigabit Ethernet line.

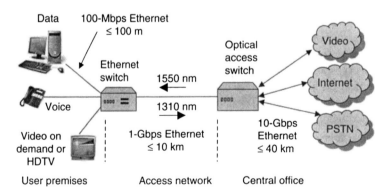

Figure 1.14. Point-to-point Ethernet optical access network.

The other option is to run either one bidirectional or two unidirectional fibers from the central office to an Ethernet switch located in the neighborhood of the subscribers. This network layout greatly reduces the number of fibers interfacing to the central office, but now the use of an Ethernet switch requires electrical power in the outside cable plant. Also, in addition to the 32 optical transceivers used for links between the subscribers and the local Ethernet switch, either two or four more optical transceivers are needed for the link running between the central office and the switch.

1.5 SUMMARY

The capacity strain that the rising demand for Internet use put on traditional twisted-pair wire links led to upgraded transmission links for access networks. Competing technologies for these new links include hybrid fiber–coax (HFC), asymmetric digital subscriber lines (ADSL), broadband wireless links, and passive optical networks (PON). Each of these technologies has strengths and limitations.

The use of a passive optical network is attractive from several viewpoints. A main driver is that since there are no active components between the endpoints of such a network, the maintenance costs are expected to be very low. Another factor is that whereas in some of the competing technologies the bandwidth that can be offered to customers decreases with increasing distance from the central office, a PON does not have such a limitation. Such advantageous factors have led to the concept of *fiber-to-the-premises* (FTTP) networks using PON technology, which is the topic of this book.

FURTHER READING

1. E. Desurvire, *Broadband Access, Optical Components and Networks, and Cryptography*, Wiley-Interscience, Hoboken, NJ, 2004.
2. W. J. Goralski, *ADSL and DSL Technologies*, McGraw-Hill, New York, 2002.
3. P. E. Green, "Fiber to the home: the next big broadband thing," *IEEE Commun. Mag.*, vol. 42, pp. 100–106, Sept. 2004.
4. A. Gumaste and T. Anthony, *First Mile Access Networks and Enabling Technologies*, Cisco Press, Indianapolis, IN, 2004.
5. IEEE Standard 802.16, "Air Interface for Fixed Broadband Wireless Access Systems," Oct. 2004.
6. G. Keiser, *Local Area Networks*, 2nd ed., McGraw-Hill, Burr Ridge, IL, 2002.
7. G. Kramer, *Ethernet Passive Optical Networks*, McGraw-Hill, New York, 2005.
8. S. Ovadia, *Broadband Cable TV Access Networks: From Technologies to Applications*, Prentice Hall, Upper Saddle River, NJ, 2001.
9. W. I. Way, *Broadband Hybrid Fiber/Coax Access System Technologies*, Academic Press, New York, 1998.

2

OPTICAL COMMUNICATIONS ESSENTIALS

In this chapter we present the fundamental concepts of optical fiber communication systems as a basis for the material in the rest of the book. Descriptions of the operating characteristics of the main components will allow the reader to understand how an optical fiber link works. These components include optical fiber types, light signal transmitters and receivers, passive optical devices, optical amplifiers, and active optical components. The emphasis in this book is on components and subsystems that are used to enable the installation of a passive optical network (PON).

After identifying the main devices in an optical fiber link, we give details on the types and performance characteristics of optical fibers, including fiber structures, light-guiding principles, and signal attenuation and dispersion. For the other link components, Chapter 4 covers optical transmitters and receivers and Chapter 5 the operation of passive optical devices used in a PON.

2.1 DEFINITIONS OF UNITS AND TERMS

To start, we look at the definitions of some units and terms that are used widely in optical communications. These include metric prefixes, electromagnetic spectral bands, optical communications spectral bands, digital signal multiplexing designations, decibels, and refractive index.

2.1.1 Metric Prefixes

A basic item that appears throughout any communications book is the prefix used in metric units for designating parameters such as length, speed, power level, and information transfer rate. Although many of these are well known, a few may be new to

---

FTTX Concepts and Applications, by Gerd Keiser
Copyright © 2006 John Wiley & Sons, Inc.

TABLE 2.1 Common Metric Prefixes, Their Symbols, and Their Magnitudes

Prefix	Symbol	Decimal	Magnitude	Multiple
yotta-	Y			10^{24}
zetta-	Z			10^{21}
exa-	E			10^{18}
peta-	P		Quadrillion	10^{15}
tera-	T		Trillion	10^{12}
giga-	G	1,000,000,000	Billion	10^{9}
mega-	M	1,000,000	Million	10^{6}
kilo-	k	1,000	Thousand	10^{3}
centi-	c	0.01	Hundredth	10^{-2}
milli-	m	0.001	Thousandth	10^{-3}
micro-	μ	0.000001	Millionth	10^{-6}
nano-	n	0.000000001	Billionth	10^{-9}
pico-	p		Trillionth	10^{-12}
femto-	f		Quadrillionth	10^{-15}
atto-	a			10^{-18}
zepto-	z			10^{-21}
yocto-	y			10^{-24}

some readers. As a handy reference, Table 2.1 lists standard prefixes, their symbols, and their magnitudes, which range in size from 10^{24} to 10^{-24}. As an example, a distance of 2×10^{-9} m (meters) = 2 nm (nanometers). The three highest and lowest designations are not especially common yet in communication systems but are included in Table 2.1 for completeness.

2.1.2 Electromagnetic Spectral Bands

All telecommunication systems use some form of electromagnetic energy to transmit signals. The *spectrum* of electromagnetic (EM) radiation is shown in Figure 2.1. *Electromagnetic energy* is a combination of electrical and magnetic fields and includes power, radio waves, microwaves, infrared light, visible light, ultraviolet

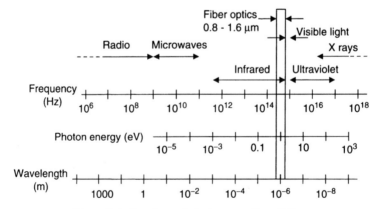

Figure 2.1. Spectrum of electromagnetic radiation.

light, x-rays, and gamma rays. Each discipline makes up a portion (or band) of the electromagnetic spectrum. The fundamental nature of all radiation within this spectrum is that it can be viewed as electromagnetic waves that travel at the speed of light, which is about $c = 300{,}000$ kilometers per second (3×10^8 m/s) or 180,000 miles per second (1.8×10^5 mi/s) in a vacuum. Note that the speed of light s in a material is less than the speed c in a vacuum, as described in Section 2.1.6.

The physical property of the waves in different parts of the spectrum can be measured in several interrelated ways: by the length of one period of the wave, by the energy contained in the wave, or by the oscillating frequency of the wave. Whereas electrical signal transmission tends to use frequency to designate the signal operating bands, optical communication generally uses *wavelength* to designate the spectral operating region and *photon energy* or *optical power* when discussing topics such as signal strength or electrooptical component performance.

As can be seen from Figure 2.1, there are three ways to measure various regions in the EM spectrum. These measurement units are related by some simple equations. First, the speed of light c is equal to the wavelength λ times the frequency ν, so $c = \lambda \nu$. Rearranging this equation gives the relationship between wavelength and frequency. For example, if the frequency is known and one wants to find the wavelength, we use

$$\lambda = \frac{c}{\nu} = \frac{3 \times 10^8 \text{ m/s}}{\nu} \qquad (2.1)$$

where the frequency ν is measured in cycles/s or hertz (Hz). Conversely, if the wavelength is known and one wants to find the frequency, one uses the relationship $\nu = c/\lambda$.

The relationship between the energy of a photon and its frequency (or wavelength) is determined by the equation known as *Planck's law*,

$$E = h\nu \qquad (2.2)$$

where the parameter $h = 6.63 \times 10^{-34}$ J·s = 4.14 eV·s is called *Planck's constant*. The unit abbreviation J represents joules, and eV, electron volts. In terms of wavelength (measured in micrometers), the energy in electron volts is given by

$$E \text{ (eV)} = \frac{1.2406}{\lambda \text{ (µm)}} \qquad (2.3)$$

2.1.3 Optical Spectral Band

The *optical spectrum* ranges from about 5 nm (ultraviolet) to 1 mm (far infrared), the visible region being the 400- to 700-nm band. Optical fiber communications use the spectral band from 800 to 1675 nm.

The International Telecommunications Union (ITU) has designated six spectral bands for use in intermediate-range and long-distance optical fiber communications within the 1260- to 1675-nm wavelength region. These band designations arose from

22 OPTICAL COMMUNICATIONS ESSENTIALS

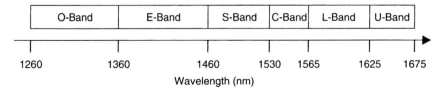

Figure 2.2. Spectral band designations for optical fiber communications.

the physical characteristics of optical fibers and the performance behavior of optical amplifiers. As shown in Figure 2.2, the regions are known by the letters O, E, S, C, L, and U, which are defined as follows:

- Original band (O-band): 1260 to 1360 nm
- Extended band (E-band): 1360 to 1460 nm
- Short band (S-band): 1460 to 1530 nm
- Conventional band (C-band): 1530 to 1565 nm
- Long band (L-band): 1565 to 1625 nm
- Ultralong band (U-band): 1625 to 1675 nm

2.1.4 Digital Multiplexing Hierarchy

When high-capacity fiber optic transmission lines started being deployed in the 1980s, service providers worldwide agreed on a standard signal format called *synchronous optical network* (SONET) in North America and *synchronous digital hierarchy* (SDH) in other parts of the world. These standards define a synchronous frame structure for sending multiplexed digital traffic over optical fiber trunk lines. The basic building block and first level of the SONET signal hierarchy is called the *synchronous transport signal—level 1* (STS-1).

Figure 2.3 shows the basic structure of an STS-1 frame. This is a two-dimensional structure consisting of 90 columns by nine rows of bytes, where 1 *byte* is 8 bits. Each

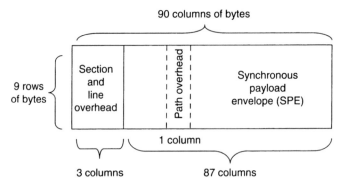

Figure 2.3. Basic structure of a SONET STS-1 frame.

TABLE 2.2 Popular SONET and SDH Transmission Rates

SONET Level	Electrical Level	SDH Level	Line Rate (Mbps)	Common Rate Name
OC-1	STS-1	—	51.84	—
OC-3	STS-3	STM-1	155.52	155 Mbps
OC-12	STS-12	STM-4	622.08	622 Mbps
OC-48	STS-48	STM-16	2,488.32	2.5 Gbps
OC-192	STS-192	STM-64	9,953.28	10 Gbps
OC-768	STS-768	STM-256	39,813.12	40 Gbps

frame is repeated 8000 times per second. Thus, each byte in the frame corresponds to a bit rate of 64 kbps. The first three columns comprise transport overhead bytes that carry network management information. The remaining field of 87 columns, called the *synchronous payload envelope* (SPE), carries 86 columns of user data plus a 9-byte column of *path overhead* (POH). The POH supports performance monitoring by the end equipment, status, signal labeling, a tracing function, and a user channel. The nine path-overhead bytes are always in a column and can be located anywhere in the SPE. Since a fundamental SONET frame is repeated 8000 times per second, it has a 125-µs duration. Thus, the transmission bit rate of the basic SONET signal is

$$\text{STS-1} = (90 \text{ bytes/row})(9 \text{ rows/frame})(8 \text{ bits/byte})(125 \text{ µs/frame})$$
$$= 51.84 \text{ Mbps}$$

Higher-rate SONET signals are obtained by byte-interleaving N of the STS-1 frames to form an STS-N frame. To have a signal pattern that conforms to the transmission characteristics of an optical fiber link, the STS-N signals are scrambled to avoid long strings of logic 1's and 0's. This action converts them to an *optical carrier—level N* (OC-N) signal, which has the same rate as the STS-N signal. Thus, the OC-N signal will have a line rate exactly N times that of a 51.84-Mbps OC-1 signal. For SDH systems the fundamental building block is the 155.52-Mbps *synchronous transport module—level 1* (STM-1). Higher-rate SDH information streams are generated by synchronously multiplexing N different STM-1 signals to form the STM-N signal. Table 2.2 shows popular SDH and SONET signal levels. The right-hand column lists the common names given to the specific rates.

2.1.5 Decibel Units

Attenuation (reduction) of the signal strength results from various loss mechanisms in a transmission medium. For example, electric power is lost through heat generation as an electric signal flows along a wire, and optical power is attenuated through scattering, reflection, and absorption processes in a glass fiber or in an atmospheric channel.

A convenient method for establishing a measure of attenuation is to reference the signal level to some absolute power value (e.g., 1 mW), to a power value in another part of the system (e.g., for loss or gain comparisons between two locations), or to a noise power level. For guided media the signal strength normally decreases exponentially, so for convenience one can designate it in terms of a logarithmic power ratio measured in decibels (dB). This is defined in terms of the base 10 logarithm by

$$\text{power ratio (dB)} = 10 \log \frac{P_2}{P_1} \tag{2.4}$$

where P_1 and P_2 are the electrical or optical power levels of a signal at points 1 and 2 in a transmission path, or P_1 can be a reference level. The logarithmic nature of the decibel allows a large ratio to be expressed in a fairly simple manner. Power levels differing by many orders of magnitude can be compared easily when they are in decibel form. Another attractive feature of the decibel is that when measuring the changes in the strength of a signal, one merely adds or subtracts the decibel numbers between two different points.

As an example, assume that after traveling a certain distance in some transmission medium, the power of a signal is reduced to half, that is, as shown in Figure 2.4, $P_2 = 0.5\, P_1$. At this point, using Eq. (2.4), the attenuation or loss of power is

$$10 \log \frac{P_2}{P_1} = 10 \log \frac{0.5 P_1}{P_1} = 10 \log 0.5 = 10(-0.3) = -3 \text{ dB}$$

Thus, -3 dB means that the signal has lost half its power. This is a 3-dB attenuation or loss. If an amplifier is inserted into the link at this point to boost the signal back to its original level, that amplifier has a 3-dB gain. If the amplifier has a 6-dB gain, it boosts the signal power level to twice the original value, and so on.

Table 2.3 shows some sample values of power loss given in decibels and the percent of power remaining after this loss. These types of numbers are important when considering factors such as the effects of tapping off a small part of an optical signal for monitoring purposes, for examining the power loss through some optical element, or when calculating the signal attenuation in a specific length of optical fiber.

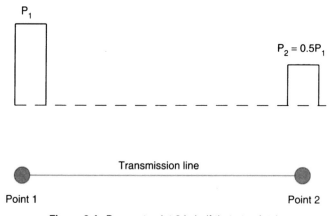

Figure 2.4. Power at point 2 is half that at point 1.

TABLE 2.3 Examples of Decibel Power Loss and the Remaining Percentages

Power Loss (dB)	Percent of Power Left
0.1	98
0.5	89
1	79
2	63
3	50
6	25
10	10
20	1

As an example of optical power loss in a transmission link, consider the transmission path from point 1 to point 4 shown in Figure 2.5. Here the signal is attenuated by 14 dB between points 1 and 2. After getting a 15-dB boost from an amplifier at point 3, it is again attenuated by 6 dB between points 3 and 4. Relative to point 1, the signal level in dB at point 4 is

decibel level at point 4 = (loss in line 1) + (amplifier gain) + (loss in line 2)
= (−14 dB) + (15 dB) + (−6 dB) = −5 dB

Thus, the signal experiences a 5-dB loss in power when it travels from point 1 to point 4. This means that it is reduced by a factor of $10^{0.5} = 3.16$.

Since the decibel is used to refer to ratios or relative units, it gives no indication of the *absolute power level*. However, a *derived* unit can be used for this. Such a unit that is particularly common in optical fiber communications is the *dBm*. This expresses the power level P as a logarithmic ratio of P referred to 1 mW. In this case, the power in dBm is an absolute value defined by

$$\text{power level (dBm)} = 10 \log \frac{P \text{ (mW)}}{1 \text{ mW}} \tag{2.5}$$

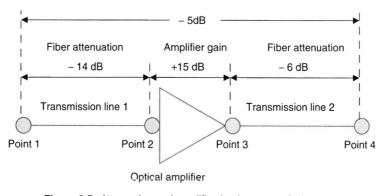

Figure 2.5. Attenuation and amplification in a transmission path.

TABLE 2.4 Examples of Optical Power Levels and Their dBm Equivalents

Power	dBm Equivalent	Power	dBm Equivalent
20 mW	13	10 μW	−20
10 mW	10	1 μW	−30
4 mW	6	100 nW	−40
2 mW	3	10 nW	−50
1 mW	0	1 nW	−60
500 μW	−3	100 pW	−70
250 μW	−6	10 pW	−80
100 μW	−10	1 pW	−90

An important *rule-of-thumb relationship* to remember for optical fiber communications is 0 dBm = 1 mW. Therefore, positive values of dBm are greater than 1 mW and negative values are less than this. Table 2.4 lists some examples of optical power levels and their dBm equivalents.

2.1.6 Refractive Index

A fundamental optical property of a material relates to how fast light travels in it. Upon entering a dielectric or nonconducting medium a lightwave slows down and now travels at a speed s, which is characteristic of the material and is less than c. The ratio of the speed of light in a vacuum to that in matter, known as the *refractive index* or *index of refraction n* of the material, is given by

$$n = \frac{c}{s} \tag{2.6}$$

Typical values of n to two decimal places are 1.00 for air, 1.33 for water, and 1.45 for silica glass. Note that if we have two different materials, the one with the larger value of n is said to be *optically denser* than the material with the lower value of n. For example, glass is optically denser than air.

2.2 ELEMENTS OF AN OPTICAL LINK

The basic function of an optical fiber link is to transport a signal from some piece of electronic equipment (e.g., a computer, telephone, or video device) at one location to corresponding equipment at another location with a high degree of reliability and accuracy. Figure 2.6 shows the key sections of an optical fiber communications link, which are as follows:

- *Optical fiber.* As described in Section 2.3, the optical fiber is one of the most important elements in an optical link. A variety of fiber types exist, and there are many different cable configurations, depending on whether the cable is to

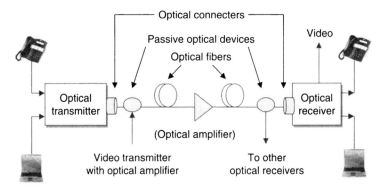

Figure 2.6. Main constituents of an optical fiber link.

be installed inside a building, in underground pipes, outside on poles, or under water (see Chapter 12).

- *Optical transmitter.* The transmitter consists of a light source and associated electronic circuitry. The source can be a light-emitting diode (LED) or a laser diode. The transmitter electronics are used to set the source operating point and to vary the optical output in proportion to an electrically formatted information input signal. Optical transmitter types and their operating characteristics are described in Chapter 4.
- *Optical receiver.* Inside the receiver is a photodiode that detects the weakened and distorted optical signal emerging from the end of an optical fiber and converts it to an electrical signal. The receiver also contains electronic amplification devices and circuitry to restore signal fidelity. Details are given in Chapter 4.
- *Passive devices.* Passive devices are optical components that require no electronic control for their operation. Among these are optical filters that select only a narrow spectrum of desired light, optical splitters that divide the power in an optical signal into a number of different branches, optical multiplexers that combine signals from two or more distinct wavelengths onto the same fiber (or that separate the wavelengths at the receiving end), and couplers used to tap off a certain percentage of light, usually for performance monitoring purposes. The functions and characteristics of passive components used in PONs are described in Chapter 5.
- *Optical amplifiers.* After an optical signal has traveled a certain distance along a fiber, it becomes greatly weakened due to power loss along the fiber. At that point the optical signal needs to get a power boost. This is done in long-distance links by means of an optical amplifier that boosts the power level completely in the optical domain. In a PON an optical amplifier is not employed in the outside cable plant but is used in a central office to boost the level of analog video signals before inserting them onto a fiber line. Chapter 10 has more details on this application.

- *Active components*. Not shown in Figure 2.6 are a wide range of other active optical components, which require electronic control for their operation. These include light signal modulators, tunable (wavelength-selectable) optical filters, variable optical attenuators, and optical switches. Since they are not used in a PON, the details on these components are not described in this book but may be found in optical fiber communication books such as those by Keiser or Freeman (see the Further Reading section).
- *Connectors and splices*. Very low-loss optical connectors and splices are needed in a PON for joining cables and for attaching one bare fiber to another. In Chapter 5 we describe some of the optical connection and splicing techniques used in PONs.

2.3 OPTICAL FIBERS

The optical fiber is a key part of a lightwave communication system. An optical fiber is a cylindrical dielectric waveguide that confines and guides light along its axis. Except for certain specialty fibers, basically all fibers used for telecommunication purposes have the same physical structure. The variations in material and size of this structure determine the transmission characteristics of a light signal and influence how the fiber responds to environmental perturbations, such as stress, bending, and temperature variations.

2.3.1 Fiber Structures

Figure 2.7 shows the end face cross section and a longitudinal cross section of a standard optical fiber, which consists of a cylindrical glass core surrounded by a glass cladding. The *core* has a refractive index n_1 and the *cladding* has a refractive index n_2. Surrounding these two layers is a polymer *buffer coating* that protects the fiber from mechanical and environmental effects. Traditionally, the core radius is designated by the letter a. In almost all cases, for telecommunication fibers the core and cladding are made of silica glass (SiO_2).

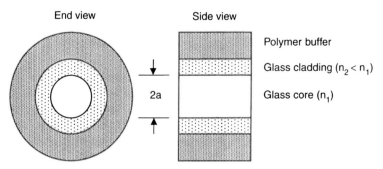

Figure 2.7. Cross sections of a generic fiber structure showing a core, a cladding, and a buffer coating.

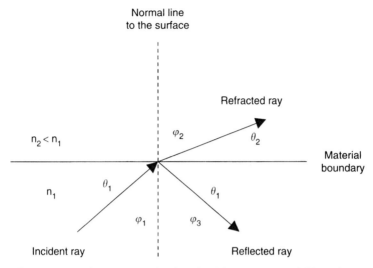

Figure 2.8. Refraction and reflection of a light ray at a material boundary.

The refractive index n_2 of the cladding is slightly smaller than the index of the core (i.e., $n_2 < n_1$), which is the condition required for light to propagate along the fiber. The difference in the core and cladding indices also determines how light signals behave within the fiber. Typically, the index differences range from 0.2 to 3.0 percent, depending on the desired behavior of the resulting fiber.

The principle of how light travels along a fiber is based on the concepts of reflection and refraction. These can be understood most easily using light rays. When a light ray encounters a boundary separating two materials that have different refractive indices, part of the ray is reflected back into the first medium and the remainder is bent (or *refracted*) as it enters the second material. This is shown in Figure 2.8. The bending or refraction of the light ray at the interface is a result of the difference in the speed of light in two materials with different refractive indices.

The relationship describing refraction at the interface between two different materials, known as *Snell's law*, is given by

$$n_1 \sin \phi_1 = n_2 \sin \phi_2 \qquad (2.7)$$

or, equivalently, as

$$n_1 \cos \theta_1 = n_2 \cos \theta_2 \qquad (2.8)$$

where the angles are as defined in Figure 2.8. The angle ϕ_1 between the incident ray and the normal to the surface is known as the *angle of incidence*. According to the *law of reflection*, illustrated in Figure 2.9, the angle ϕ_1 at which the incident ray strikes the interface is exactly equal to the angle ϕ_3 that the reflected ray makes with the same interface.

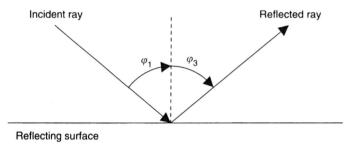

Figure 2.9. Law of reflection.

When light traveling in a certain medium is reflected off an optically denser material (one with a higher refractive index), the process is referred to as *external reflection*. Conversely, the reflection of light off less optically dense material (such as light traveling in glass being reflected at a glass-to-air interface) is called *internal reflection*.

As the angle of incidence φ_1 in an optically denser material becomes larger, the refracted angle φ_2 approaches $\pi/2$. Beyond this point no refraction into the adjoining material is possible and the light rays become *totally internally reflected*. The conditions required for total internal reflection can be determined by using Snell's law. Consider Figure 2.10, which shows a glass surface in air. A light ray gets bent toward the glass surface as it leaves the glass in accordance with Snell's law. If the angle of incidence φ_1 is increased, eventually a point will be reached where the light ray in air is parallel to the glass surface. The center diagram shows this event. This point is known as the *critical angle of incidence* φ_c. When φ_1 is greater than φ_c, the condition for total internal reflection is satisfied; that is, the light is totally reflected back into the glass with no light escaping from the glass surface.

To get an understanding of how much light a fiber accepts, let us examine the case when the core diameter is much larger than the wavelength of the light. For such a case we can consider a simple geometric optics approach using the concept of light rays. Figure 2.11 shows a light ray entering the fiber core from a medium of refractive index n, which is less than the index n_1 of the core. The ray meets the core end face at an angle θ_0 with respect to the fiber axis and is refracted into the core. Inside

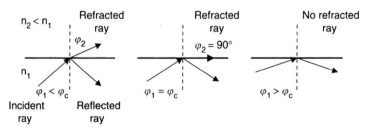

Figure 2.10. Representation of the critical angles and total internal reflection at a glass-to-air interface.

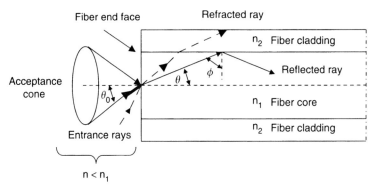

Figure 2.11. Ray optics representation of the propagation mechanism in an ideal step-index optical waveguide

the core the ray strikes the core–cladding interface at a normal angle ϕ. If the light ray strikes this interface at such an angle that it is totally reflected, the ray follows a zigzag path along the fiber core.

Now suppose that the angle θ_0 is the largest entrance angle for which total internal reflection can occur at the core–cladding interface. Then rays outside the acceptance cone shown in Figure 2.11, such as the ray given by the dashed line, will refract out of the core and be lost in the cladding. This condition defines a *critical angle* ϕ_c, which is the smallest angle ϕ that supports total internal reflection at the core–cladding interface.

Reflection of some light at a fiber end is an important point to consider when discussing optical power coupling between a light source and a fiber or between two optical fibers that have a gap between them. If light traveling in a medium of refractive index n_1 is incident perpendicularly on another medium of index n_2, the power coupled into the second medium is reduced by the factor

$$R = \left(\frac{n_2 - n_1}{n_2 + n_1}\right)^2 \tag{2.9}$$

where R is the *Fresnel reflection* or *reflectivity* at the fiber end face. The *reflection coefficient*, which is given by the ratio $r = (n_2 - n_1)/(n_2 + n_1)$, relates the amplitude of the reflected wave to the amplitude of the incident wave.

2.3.2 Rays and Modes

Although it is not directly obvious from the simple ray picture shown in Figure 2.11, only a finite set of rays at certain discrete angles are capable of propagating along a fiber. These angles are related to a set of electromagnetic wave patterns or field distributions called *modes* that can propagate along a fiber. When the fiber core diameter is on the order of 8 to 10 μm, which is only a few times the value of the wavelength, only the single *fundamental ray* that travels straight along the fiber axis is allowed to propagate. Such a fiber is referred to as a *single-mode fiber*. Fibers with larger core diameters (e.g., 50 or 62.5 μm) support many propagating rays or modes and are known as *multimode fibers*.

The electromagnetic field patterns and the corresponding optical power distributions of the modes in an optical fiber are rather complex, and various simplified models have been developed to approximate them. However, one important characteristic of these modes is that their power distributions are not confined completely to the core, but instead, extend partially into the cladding. The fields vary harmonically within the core guiding region of index n_1 and decay exponentially outside this region (in the cladding). Whereas for low-order modes the fields are concentrated tightly near the axis of the fiber with little penetration into the cladding, for higher-order modes the fields are distributed more toward the edges of the core and penetrate farther into the cladding region.

As the simplest example, consider the approximation of the relative optical power distribution of the lowest-order mode shown in Figure 2.12. This corresponds to the fundamental ray and is the only mode that can propagate in a single-mode fiber. As Figure 2.12 shows, optical power is concentrated at the center of the fiber core, has a bell-shaped envelope, and decays exponentially in the cladding. With respect to the peak power, the level at the core–cladding interface is about 20 percent. The size of the mode is sensitive to wavelength; that is, the mode is confined more to the core as the wavelength decreases.

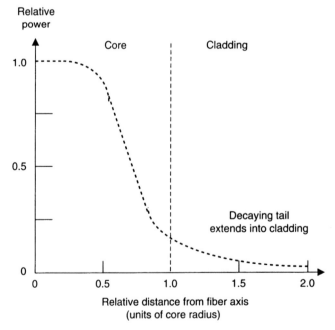

Figure 2.12. Optical power distribution in the core and cladding of the lowest-order guided mode of an optical fiber.

The characteristic that the power of a mode extends partially into the cladding is of importance in applications such as coupling of power from one fiber to another. For example, this principle is used in designing the fused-fiber optical couplers described in Chapter 3.

2.4 OPTICAL FIBER ATTENUATION

Light traveling in a fiber loses power over distance, mainly because of absorption and scattering mechanisms in the fiber. The fiber loss is referred to as *signal attenuation* or simply *attenuation*. This is an important property of an optical fiber because together with signal distortion mechanisms, attenuation determines the maximum transmission distance possible between a transmitter and a receiver in a PON without intermediate amplification. The optical signal power at the receiver must have an appropriate level above the signal noise for high-fidelity reception. The degree of the attenuation depends on the wavelength of the light and on the fiber material.

Figure 2.13 shows a generic attenuation-versus-wavelength curve for a silica fiber. The total loss of power is measured in decibels, and the loss per length within a cable is described in decibels per kilometer (dB/km). As an example, suppose that a fiber has an attenuation of 0.5 dB/km at a 1310-nm wavelength. Then after traveling 20 km, the optical power loss in the fiber is 10 dB (a factor of 10). If the attenuation at 1550 nm is 0.3 dB/km, the power loss over a 20-km fiber is 6 dB (a factor of 4).

Light power also can be lost as a result of fiber bending. Fibers can be subject to two types of bends: (1) *macroscopic bends*, which have radii that are large compared

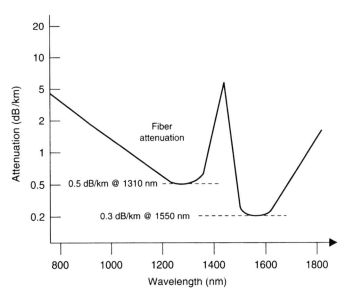

Figure 2.13. Generic attenuation-versus-wavelength curve for a silica optical fiber.

34 OPTICAL COMMUNICATIONS ESSENTIALS

with the fiber diameter, for example, such as those that occur when a fiber cable turns a sharp corner, for example, in an equipment rack; and (2) random *microscopic bends* of the fiber axis, which can arise when fibers are incorporated into cables. Since the microscopic bending loss is determined in the manufacturing process, the user has little control over the degree of loss resulting from them. In general, cable fabrication processes keep these values to a very low value, which is included in published cable-loss specifications.

For slight bends, the excess optical power loss due to macroscopic bending is extremely small and is essentially unobservable. As the radius of curvature decreases, the loss increases exponentially until at a certain critical bend radius the loss becomes observable. If the bend radius is made a bit smaller once this threshold has been reached, the losses suddenly become extremely large. Bending losses depend on wavelength and are greater at longer wavelengths. As a rule of thumb, it is best not to make the bend radius of a standard telecommunication fiber be less than 2.5 cm.

The need to adhere to maximum bending rules is especially important when threading a fiber patch cord through an equipment rack where tight bends could occur. As Figure 2.14 illustrates, suppose that a 2-m optical fiber patch cord connects circuit cards in two different shelves of a rack. Here, overly tight bends might occur inadvertently at the edges of the rack or within compact coils of excess fiber length.

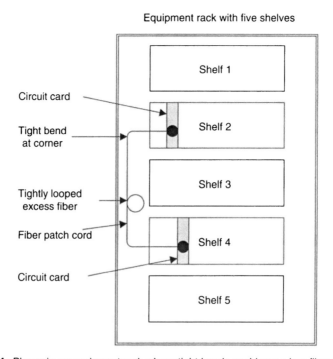

Figure 2.14. Places in an equipment rack where tight bends could occur in a fiber patch cord.

Since fibers often need to be bent into very tight loops within component packages, special fibers that are immune to bending losses have been developed for such applications. For example, such bend-insensitive fibers incur additional losses of less than 0.001 dB when they are coiled into spools with a 13-mm radius.

2.5 FIBER INFORMATION CAPACITY

The information-carrying capacity is limited by various signal dispersion factors in the optical fiber. The three main dispersion categories are modal, chromatic, and polarization mode dispersions. These distortion mechanisms cause optical signal pulses to broaden as they travel along a fiber. Figure 2.15 shows that as optical pulses travel in a fiber, they broaden due to dispersion and decrease in power because of attenuation. Thus, if they travel sufficiently far, they eventually will overlap with neighboring pulses, thereby creating errors in the output since they become indistinguishable to the receiver.

2.5.1 Modal Dispersion

Modal dispersion arises from the different path lengths associated with various modes (as represented by light rays at different angles). It appears only in multimode fibers, since in a single-mode fiber there is only one mode. Looking at Figure 2.16, it can be deduced that rays bouncing off the core–cladding interface follow a longer path than that of the fundamental ray, which travels straight down the fiber axis. For

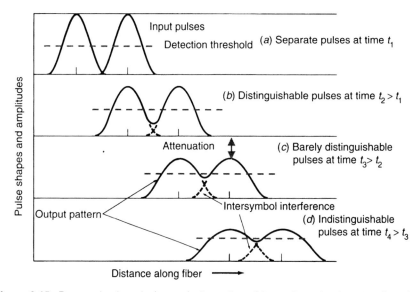

Figure 2.15. Progressive broadening and attenuation of two adjacent pulses traveling along a fiber.

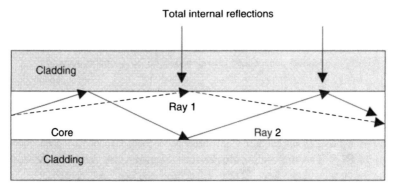

Figure 2.16. Light rays with steep incident angles have longer path lengths than lower-angle rays.

example, since ray 2 makes a steeper angle than ray 1, it has a longer path length from the beginning to the end of a fiber. If all the rays are launched into a fiber at the same time in a given light pulse, they will arrive at the fiber end at slightly different times. This causes the pulse to spread out and is the basis of modal dispersion.

2.5.2 Chromatic Dispersion

The index of refraction of silica varies with wavelength; for example, it ranges from 1.453 at 850 nm to 1.445 at 1550 nm. In addition, a light pulse from an optical source contains a certain slice of wavelength spectrum. For example, a laser diode source may emit pulses that have most of the optical power contained within a 1-nm spectral width. Consequently, different wavelengths within an optical pulse travel at slightly different speeds through the fiber (recall that the speed $s = c/n$). Therefore, each wavelength will arrive at the fiber end at a slightly different time, which leads to pulse spreading. This factor, called *chromatic dispersion* (CD), is often referred to simply as *dispersion*.

Dispersion is a fixed quantity at a specific wavelength and is measured in units of picoseconds per kilometer of fiber per nanometer of optical source spectral width, abbreviated as ps/(km·nm). For example, a single-mode fiber might have a chromatic dispersion value of $D_{CD} = 2$ ps/(km·nm) at 1550 nm. Figure 2.17 shows the approximate chromatic dispersion as a function of wavelength for several different standard fiber types, which are described in Section 2.7.

As a result of chromatic dispersion effects, an optical pulse in which the light occupies a spectral width of $\Delta\lambda$ broadens after a distance L by an amount $\Delta T_{CD} = L D_{CD} \Delta\lambda$. This broadening must be kept small enough so that neighboring signal pulses do not overlap to the point where they become indistinguishable from each other. At a data rate of B bits per second, a sufficient criterion for this condition to be satisfied is $B \Delta T < 1$.

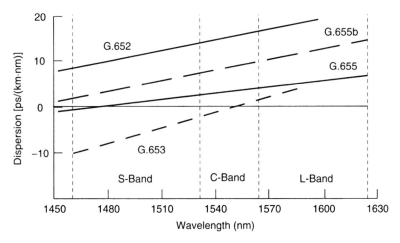

Figure 2.17. Estimated curves of chromatic dispersion as a function of wavelength for several different fiber types.

2.5.3 Polarization Mode Dispersion

The *polarization mode dispersion* effect, commonly called PMD, results from the fact that light-signal energy at a given wavelength in a single-mode fiber actually occupies two orthogonal polarization states or modes. Figure 2.18 shows this condition. At the start of the fiber the two polarization states are aligned. However, the refractive index of optical fiber material is not perfectly uniform across any given cross-sectional area and varies along the length of the fiber. This condition is known as *birefringence* of the material. Consequently, each polarization mode will encounter a slightly different refractive index, so that one polarization mode will travel faster

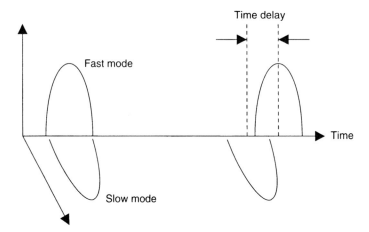

Figure 2.18. Variation in the polarization states of an optical pulse as it passes through a fiber that has varying birefringence along its length.

than the other. The resulting difference in propagation times between the two orthogonal polarization modes causes pulse spreading. This is the basis of polarization mode dispersion.

PMD is not a fixed quantity but fluctuates with time due to factors such as temperature variations and stress changes on the fiber. It varies as the square root of distance and thus is specified as a mean value in units of ps/$\sqrt{\text{km}}$. A typical value is $D_{PMD} = 0.05$ ps/$\sqrt{\text{km}}$. As a result of PMD effects, an optical pulse broadens after a distance L by an amount $\Delta T_{PMD} = D_{PMD} L^{1/2}$.

2.6 NONLINEAR EFFECTS IN FIBERS

Nonlinear effects occur when there are high light power densities (optical power per cross-sectional area) in a fiber. Their impact on signal fidelity includes backscattering of some fraction of the signal power, shifting of power between wavelength channels, appearances of spurious signals at other wavelengths, and decreases in signal strength.

Nonlinear effects can be classified into two categories. The first encompasses the nonlinear *inelastic scattering processes* known as *stimulated Raman scattering* (SRS) and *stimulated Brillouin scattering* (SBS). These two processes have the most significant effect in a PON in relation to the quality of analog video signals.

The second category of nonlinear effects arises from *intensity-dependent variations* in the refractive index in a silica fiber. This produces effects such as *self-phase modulation* (SPM), *cross-phase modulation* (XPM), and *four-wave mixing* (FWM). The processes SBS, SRS, and FWM result in gains or losses in a wavelength channel that are dependent on the optical signal intensity. These nonlinear processes provide gains to some channels while depleting power from others, thereby producing crosstalk between the wavelength channels.

Both SPM and XPM affect only the phase of signals, which causes chirping in digital pulses. This can worsen pulse broadening due to dispersion, particularly in very high-rate systems (>10 Gbps). These two nonlinear processes are not a major concern for PONs.

2.6.1 Stimulated Brillouin Scattering

Stimulated Brillouin scattering arises from the fact that a strong optical signal propagating in a fiber generates an acoustic wave that produces variations in the refractive index. These index variations cause lightwaves to scatter in the backward direction toward the transmitter. The effect of SBS is negligible for low power levels, but becomes greater as the optical power level increases. At a certain signal level threshold the SBS process becomes nonlinear and the launched signal loses an increasingly greater percentage of power. Figure 2.19 shows this phenomenon in terms of the relative propagating signal level and the backscattered power after traveling a certain distance along a fiber. Beyond the SBS threshold the percentage increase in signal depletion continues to grow until the SBS limit is reached. Any

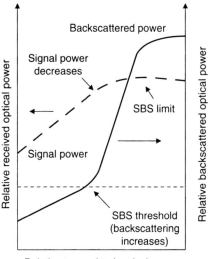

Figure 2.19. Effect of SBS on signal power in an optical fiber.

additional optical power launched into the fiber after this point merely is scattered backward along the fiber due to the SBS effect.

Standard G.652 single-mode fiber has an SBS limit of 17 dBm. As an example of efforts to mitigate this limit, Corning Inc. designed a G.652-compatible fiber with a 3-dB-higher SBS threshold. This fiber enables an analog video signal with twice as much power (an additional 3 dB) to be launched into the fiber.

2.6.2 Stimulated Raman Scattering

Stimulated Raman scattering is an interaction between photons in an optical signal and the vibrational modes of silica molecules. In this interaction a molecule can absorb some energy from the photon, which causes the photon to attain a lower energy. The modified photon is called a *Stokes photon*. Since energy is inversely proportional to wavelength [see Eq. (2.3)], this means that the Stokes photon has a higher wavelength. Because the optical signal wave that is injected into a fiber is the source of the interaction, it is called the *pump wave*, since it supplies power for the newly generated wave.

Thus, the SRS process generates scattered light at a wavelength longer than that of the incident light. If another signal is present at this longer wavelength, the SRS light will amplify it and the pump-wavelength signal will decrease in power. Figure 2.20 illustrates the SRS effect, which occurs over a wide wavelength range. Powers in channels separated by up to 16 THz (125 nm) can be coupled through the SRS effect, thereby producing crosstalk between wavelength channels. As shown in Figure 2.20, for PON applications the 1490-nm signal acts as a pump for the 1550 nm video signal, since it is only 50 nm away. However, if the optical power at 1490 nm

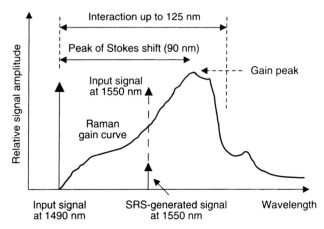

Figure 2.20. SRS gain curve. Here a 1490-nm signal acts as a pump for a 1550-nm signal.

is less than about 2 mW, the effects of SRS-induced interference in the video signal are small.

2.7 OPTICAL FIBER STANDARDS

The Telecommunications Sector of the International Telecommunications Union (ITU-T) and the Telecommunications Industry Association in conjunction with the Electronics Industries Alliance (TIA/EIA) are the main organizations that have published standards for both multimode and single-mode optical fibers used in telecommunications. The recommended bounds on fiber parameters (e.g., attenuation, cutoff wavelength, and chromatic dispersion) designated in these standards assure the users of product performance consistency. In addition, the standards allow fiber manufacturers to have a reasonable degree of flexibility to improve products and develop new ones.

Multimode fibers typically are used for local area network (LAN) applications in a campus environment, particularly for gigabit or 10-gigabit rate Ethernet links, which are known popularly as GigE and 10GigE, respectively. They also can be used within the customer's premises in a PON. Applications include links in an office or government building, a medical facility, or a manufacturing plant, where the distance between buildings is typically 300 m or less. The two principal multimode fiber types for these applications have either 50- or 62.5-μm core diameters and both have 125-μm cladding diameters.

TABLE 2.5 Operating Ranges of Various Multimode Fibers for Applications Up to 10GigE

Parameter	62.5-μm Fiber	50-μm Fiber	Unit
Bandwidth	160 to 200	400 to 3000	MHz·km
Range	26 to 33	62 to 300	m
Attenuation	2.5	2.5	dB/km

TABLE 2.6 Operating Parameters of Typical Single-Mode Fibers at 1310 and 1550 nm

Parameter	Fiber at 1310 nm	Fiber at 1550 nm	Unit
Attenuation (max.)	0.40	0.30	dB/km
Mode field diameter	9.3 ± 0.5	10.5 ± 1.0	µm
Macrobending loss	≤0.05	≤0.10	dB
Cabled PMD	≤0.05	≤0.05	ps/$\sqrt{km}$
Fiber weight	64	64	g/km

TABLE 2.7 ITU-T Recommendations for Telecommunication Optical Fibers

ITU-T Rec. No.	Description
G.651	Multimode fiber for use at 850 nm in a campus environment
G.652a, b	Standard single-mode fiber (1310-nm optimized)
G.652c	Low-water-peak fiber for coarse WDM applications (see Chapter 3)
G.652d	Low-water-peak and low-dispersion fiber for PON applications
G.653	Dispersion-shifted fiber (made obsolete by G.655 fiber)
G.654	Used mainly for long-distance undersea applications
G.655a, b	Nonzero dispersion-shifted fiber (NZDSF) for long-haul links
G.656	Low chromatic dispersion fiber for CWDM applications

Table 2.5 shows the operating ranges of various multimode fibers for applications up to 10GigE. The standards document TIA/EIA-568 lists the specifications for 10GigE fiber. The ITU-T Recommendation G.651 describes other multimode fiber specifications for LAN applications using 850-nm optical sources.

Table 2.6 shows operating parameter values of single-mode fibers at 1310 and 1550 nm. Here the mode field diameter specifies the light acceptance capability and is very close in size to the fiber core diameter. The macrobending specification is for the increase in loss when 100 turns of the fiber are wound on a 75-mm (3-in.)-diameter rod. These types of fibers are used in the outside plant in passive optical networks and within a central office for connecting optical transmission equipment to the incoming fiber cable.

The ITU-T has a series of recommendations for multimode and single-mode fibers. The characteristics of these fibers are summarized in Table 2.7. Various grades of single-mode G.652 fibers are used for PON applications. The other fiber types are mainly for metro and long-distance telecommunication applications.

2.8 SUMMARY

This chapter provides some fundamental operational concepts and definitions of terminology related to passive optical networks. Some key concepts include the following:

- In Section 2.1 we give basic definitions of metric prefixes, illustrate electromagnetic and optical spectral bands, describe the SONET and SDH digital signal

multiplexing hierarchy, and define the concepts of decibels and the relative power unit dBm.
- The key sections and components of a PON, which include optical fibers, transmitters, receivers, and various passive optical devices, are described in Section 2.2.
- In addressing optical fibers, in Section 2.3 we discuss their physical structure, the principle of how light travels along a fiber, and the concepts of light rays and modes.
- Attenuation and signal loss are important parameters that need careful consideration when designing an optical network. This is the topic of Section 2.4, which includes factors such as inherent fiber attenuation and bending losses.
- The factors that limit the information-carrying capacity of optical fibers are described in Sections 2.5 and 2.6. These sections address signal dispersion categories and nonlinear optical effects, respectively.
- For optical fibers the ITU-T and the TIA/EIA are the main organizations that have created and published standards for telecommunication applications. In Section 2.7 we describe briefly some of these standards. More details on standards for PON equipment are provided in subsequent chapters.

PROBLEMS

2.1 (a) What are the energies in electron volts (eV) of light at wavelengths 850, 1310, 1490, and 1550 nm?

(b) Consider a 1-ns pulse with a 100-nW amplitude at each of these wavelengths. How many photons are in such a pulse at each wavelength?

2.2 To insert low-speed signals such as 64-kbps voice channels into a SONET frame, 84 columns in each SPE are divided into seven groups of 12 columns. Each such group is called a *virtual tributary*.

(a) What is the bit rate of such a virtual tributary?

(b) How many 64-kbps voice channels can a virtual tributary accommodate?

(c) What is the payload efficiency?

2.3 Assume that the attenuation of an optical fiber is 0.4 dB/km at 1310 nm. If 200 µW of optical power is launched into the fiber, what is the output power level at the end of a 20-km link?

2.4 (a) Verify the numbers for the percentage of power remaining following the optical power losses listed in Table 2.3.

(b) Verify the unit conversions from optical power in watts to dBm listed in Table 2.4.

2.5 Light traveling in air strikes a glass surface at an angle $\theta_1 = 33°$, where θ_1 is measured between the incoming ray and the glass surface. At this surface, part of the beam is reflected and part is refracted as it enters the glass.

(a) If the refracted and reflected rays make a 90° angle with each other, what is the refractive index of the glass?

(b) What is the critical angle for this glass?

2.6 A point source of light is 12 cm below the surface of a large body of water for which $n = 1.33$. What is the radius of the largest circle on the water surface through which the light can emerge?

2.7 A 45°–90°–45° prism is immersed in a liquid that has a refractive index $n = 1.45$. What is the minimum refractive index the prism must have if a ray incident normally on one of the short faces is to be totally reflected at the long face of the prism?

2.8 A certain optical fiber has an attenuation of 0.6 dB/km at 1310 nm and 0.3 dB/km at 1550 nm. Suppose that the following two pulses are launched simultaneously into the fiber: an optical power of 150 µW at 1310 nm and an optical power of 100 µW at 1550 nm. What are the power levels in microwatts of these signals at **(a)** 8 km; **(b)** 20 km?

2.9 An optical signal has lost 60 percent of its power after traversing 4 km of fiber. What is the attenuation in dB/km of this fiber?

2.10 A continuous 15-km optical fiber link has a loss of 0.5 dB/km.

(a) What is the minimum optical power level that must be launched into the fiber to have a level of 300 nW received at the end?

(b) What input power is required for such a level if the fiber has a loss of 0.35 dB/km?

2.11 The chromatic dispersion for a G.652 optical fiber can be represented by the expression

$$D(\lambda) = \frac{\lambda S_0}{4}\left[1 - \left(\frac{\lambda_0}{\lambda}\right)^4\right]$$

Here the parameter S_0 is the value of the *dispersion slope* $S(\lambda) = dD_{CD}/d\lambda$ at the wavelength λ_0. If $S_0 = 0.0970$ ps/(nm²·km) at $\lambda_0 = 1310$ nm, plot the chromatic dispersion in the wavelength range $1270 \leq \lambda \leq 1340$ nm.

2.12 A typical G.653 fiber has a zero-dispersion wavelength at $\lambda_0 = 1550$ nm with a dispersion slope of $S_0 = 0.070$ ps/(nm²·km). For such a fiber the chromatic dispersion can be found from the expression $D(\lambda) = (\lambda - \lambda_0) S_0$.

(a) Plot the chromatic dispersion in the wavelength range $1500 \leq \lambda \leq 1600$ nm.

(b) Compare the chromatic dispersion at 1500 nm with the dispersion value for the G.652 fiber described in Problem 2.11.

2.13 Suppose that the maximum allowed pulse spreading due to PMD is 20 percent of a bit period.

(a) What is the maximum allowed spreading for bit rates of 1, 2.5, and 10 Gbps? Express the results in picoseconds.

(b) If $D_{PMD} = 0.05\,\text{ps}/\sqrt{\text{km}}$, over what distance can signals at each of these rates be transmitted?

FURTHER READING

1. J. A. Buck, *Fundamentals of Optical Fibers*, Wiley, New York, 2004.
2. A. B. Carlson, P. Crilly, and J. Rutledge, *Communication Systems*, 4th ed., McGraw-Hill, Burr Ridge, IL, 2002.
3. F. Forghieri, R. W. Tkach, and A. R. Chraplyvy, "Fiber nonlinearities and their impact on transmission systems," Chap. 8, pp. 196–264, in I. P. Kaminow and T. L. Koch, eds., *Optical Fiber Telecommunications: III*, Vol. A, Academic Press, New York, 1997.
4. R. Freeman, *Fiber Optic Systems for Telecommunications*, Wiley, New York, 2002.
5. G. Keiser, *Optical Communications Essentials*, McGraw-Hill, New York, 2003.
6. G. Keiser, *Optical Fiber Communications*, 3rd ed., McGraw-Hill, New York, 2000.
7. G. Mahlke and P. Gössing, *Fiber Optic Cables: Fundamentals, Cable Design, System Planning*, 4th ed., Wiley, New York, 2001.
8. R. Ramaswami and K. N. Sivarajan, *Optical Networks*, 2nd ed., Morgan Kaufmann, San Francisco, CA, 2002.

3

WAVELENGTH-DIVISION MULTIPLEXING

An attractive aspect of an optical communication link is that many different independent optical information channels can be sent along a single fiber simultaneously. This is done by selecting narrow-wavelength bands from the spectral region ranging across the O-band through the L-band. The technology of combining a number of information-carrying wavelengths onto the same fiber is known as *wavelength-division multiplexing* (WDM).

A major point to keep in mind when implementing a WDM system is that the wavelengths (or optical frequencies) in WDM must be spaced properly to avoid interference between adjacent channels. A basic PON design accommodates this through the selection of wavelengths at 1310, 1490, and 1550 nm. These large spacings allow the peak operating wavelengths of the end components to drift widely with factors such as large temperature changes and manufacturing-related component variations. This enables the designers to use inexpensive transceivers that do not require stringent wavelength control.

The key system features of WDM are the following:

- *Capacity upgrade*. The classical application of WDM has been to upgrade the capacity of existing point-to-point fiber optic transmission links. If each wavelength supports an independent network channel of a few gigabits per second, WDM can increase the capacity of a fiber system dramatically with each additional wavelength channel. This is a feature that can allow future upgrades to basic PON implementations.
- *Transparency*. An important aspect of WDM is that each optical channel can carry any transmission format. By using different wavelengths, fast or slow asynchronous and synchronous digital data and analog information can be sent simultaneously, and independently, over the same fiber without the need for

---

FTTX Concepts and Applications, by Gerd Keiser
Copyright © 2006 John Wiley & Sons, Inc.

a common signal structure. This is an important feature for triple-play (simultaneous voice, data, and video transmissions) PON implementations.
- *Bidirectional transmission.* Independent wavelength channels can be sent in either direction on the same fiber. For example, in basic PON applications the wavelengths 1490 and 1550 nm are sent on one fiber from the central office to the user and a 1310-nm channel is used in the reverse direction on the same fiber.
- *Wavelength routing.* Instead of using electronic means to switch optical signals at a node, a wavelength-routing network can provide a pure optical end-to-end connection between users. This is done by means of *lightpaths* that are routed and switched at intermediate nodes in the network. In some cases, lightpaths may be converted from one wavelength to another wavelength along their route. Normally, sophisticated versions of this function are not used for PONs. However, the use of optical couplers to combine and separate different wavelengths onto and off a single fiber in a PON system is a simple passive wavelength routing mechanism.

In this chapter we first address the operating principles of WDM, the spectral regions that are used for its implementation, and the functions of a generic WDM link. The next topics include the standardized spectral grids for two different WDM schemes, the spectral operating regions designated for PONs, the types of WDM couplers used for PONs, and the operational concepts of bidirectional links.

The details of different passive optical components used for multiwavelength PONs are given in Chapter 5. In Chapter 6 we show how all the different components fit together to form completely passive networks that have no active components between the transmitting and receiving ends. Chapter 11, on PON designs, covers how the components are selected and arranged to meet network performance requirements.

3.1 OPERATIONAL PRINCIPLES OF WDM

Since the spectral width of a high-quality source occupies only a narrow slice of optical bandwidth, there are many independent operating regions across the spectrum, ranging from the O-band through the L-band, that can be used simultaneously. The original use of WDM was to upgrade the capacity of installed point-to-point transmission links. This was achieved with wavelengths that were separated from several tens up to 200 nm in order not to impose strict wavelength-tolerance requirements on the different laser sources and the receiving wavelength splitters.

Subsequently, the development of lasers that have extremely narrow spectral-emission widths allowed wavelengths to be spaced less than a nanometer apart. This is the basis of wavelength-division multiplexing, which simultaneously uses a number of light sources, each emitting at a slightly different peak wavelength. Each wavelength carries an independent signal, so that the link capacity is increased greatly. The main trick is to ensure that the peak wavelength of a source is spaced sufficiently far from its neighbor so as not to create interference between their spectral extents. Equally important is the requirement that during the operation of a system these peak wavelengths do not drift into the spectral territory occupied by adjacent channels. In addition to maintaining strict control of the wavelength, system designers include an empty guardband between the channels as an operations safety factor. Thereby the fidelities of the independent messages from each source are maintained for subsequent conversion to electrical signals at the receiving end.

3.1.1 WDM Operating Regions

The possibility of having an extremely high-capacity link by means of WDM can be seen by examining the characteristics of a high-quality optical source. As an example, a distributed-feedback (DFB) laser has a frequency spectrum on the order of 1 MHz, which is equivalent to a spectral linewidth of 10^{-5} nm (see Chapter 4). With such spectral widths, simplex systems make use of only a tiny portion of the transmission bandwidth capability of a fiber. This can be seen from Figure 3.1, which depicts the attenuation of light in a silica fiber as a function of wavelength. The curve shows that the two low-loss regions of a standard G.652 single-mode fiber extend over the O-band wavelengths ranging from about 1270 to 1350 nm (originally called the *second window*) and from 1480 to 1600 nm (originally called the *third window*).

We can view these regions either in terms of *spectral width* (the wavelength band occupied by the light signal) or by means of *optical bandwidth* (the frequency band

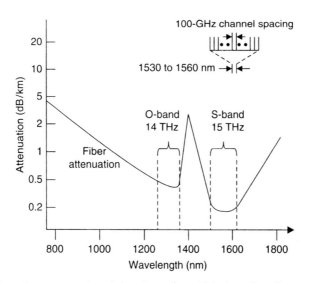

Figure 3.1. Generic representation of the attenuation of light in a silica fiber as a function of wavelength.

occupied by the light signal). To find the optical bandwidth corresponding to a particular spectral width in these regions, we use the fundamental relationship $c = \lambda \nu$, which relates the wavelength λ to the carrier frequency ν, where c is the speed of light. Differentiating this, we have

$$\Delta \nu = \frac{c}{\lambda^2} \Delta \lambda \qquad (3.1)$$

where the frequency deviation $\Delta \nu$ corresponds to the wavelength deviation $\Delta \lambda$ around λ.

Now suppose we have a fiber that has the attenuation characteristic shown in Figure 3.1. From Eq. (3.1) the optical bandwidth is $\Delta \nu = 14$ THz for a usable spectral band $\Delta \lambda = 80$ nm in the center of the O-band. Similarly, $\Delta \nu = 15$ THz for a usable spectral band $\Delta \lambda = 120$ nm in the low-loss region running from near the beginning of the S-band to almost the end of the L-band. This yields a total available fiber bandwidth of about 30 THz in the two low-loss windows.

Prior to about 2000, the peak wavelengths of adjacent light sources typically were restricted to be separated by 0.8 to 1.6 nm (100 to 200 GHz) in a WDM system. This was done to take into account possible drifts of the peak wavelength due to aging or temperature effects, and to give both the manufacturer and the user some leeway in specifying and choosing the precise peak emission wavelength. The next generation of WDM systems specified both narrower and much wider channel spacings depending on the application and on the wavelength region being used. The much narrower spacings thus require strict wavelength control of the optical source. On the other hand, the wider wavelength separations offer inexpensive WDM implementations since wavelength control requirements are relaxed significantly.

3.1.2 Generic WDM Link

The implementation of WDM networks requires a variety of passive and/or active devices to combine, distribute, isolate, add, drop, attenuate, and amplify optical power at different wavelengths. Passive devices require no external electric power or control for their operation, so they have a fixed application in WDM networks. These passive components are used to separate and combine wavelength channels, to divide optical power onto a number of fiber lines, or to tap off part of an optical signal for monitoring purposes. The performance of active devices can be controlled electronically, thereby providing a large degree of network flexibility. Active WDM components include tunable optical filters, tunable light sources, configurable add/drop multiplexers, dynamic gain equalizers, and optical amplifiers.

Figure 3.2 shows the implementation of a simple WDM link. The transmitting side has a series of independently modulated fixed-wavelength light sources, each of which emits signals at a unique wavelength. Here a *multiplexer* (popularly called a *mux*) is needed to combine these optical outputs into a continuous spectrum of signals and couple them onto a single fiber. Within a standard telecommunication link there may be various types of optical amplifiers, a variety of specialized active

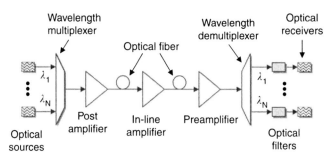

Figure 3.2. Implementation of a simple WDM link.

components (not shown), and passive optical power splitters. The operations and maintenance benefits of PONs are that no active devices are used between the transmitting and receiving endpoints.

At the receiving end a *demultiplexer* is required to separate the individual wavelengths of the independent optical signals into appropriate detection channels for signal processing. At the transmitter the basic design challenge is to have the multiplexer provide a low-loss path from each optical source to the multiplexer output. A different requirement exists for the demultiplexer, since photodetectors usually are sensitive over a broad range of wavelengths, which could include all the WDM channels. To prevent spurious signals from entering a receiving channel, that is, to give good channel isolation of the different wavelengths being used, the demultiplexer must exhibit narrow spectral operation or very stable optical filters with sharp wavelength cutoffs must be used. The tolerable crosstalk levels between channels can vary widely depending on the application. In general, a $-10\,\text{dB}$ level is not sufficient, whereas a level of $-30\,\text{dB}$ is acceptable.

In principle, any optical demultiplexer can also be used as a multiplexer. For simplicity, the word *multiplexer* is used as a general term to refer to both combining and separating functions, except when it is necessary to distinguish the two devices or functions.

3.2 STANDARD WDM SPECTRAL GRIDS

Since WDM is essentially frequency-division multiplexing at optical carrier frequencies, the WDM standards developed by the Telecommunication Sector of the International Telecommunication Union (ITU-T) specify channel spacings in terms of frequency. The first ITU-T specification for WDM was Recommendation G.692, *Optical Interfaces for Multichannel Systems with Optical Amplifiers*. This document specifies selecting the channels from a grid of frequencies referenced to 193.100 THz (1552.524 nm) and spacing them 100 GHz (about 0.8 nm at 1550 nm) apart. Suggested alternative spacings in G.692 include 50 and 200 GHz, which correspond to spectral widths of 0.4 and 1.6 nm, respectively, at 1550 nm.

TABLE 3.1 Sample Portion of the ITU-T G.694.1 DWDM Grid for 100- and 50-GHz Spacings in the L- and C-Bands

Unit	L-Band				C-Band			
	100-GHz		50-GHz Offset		100-GHz		50-GHz Offset	
	THz	nm	THz	nm	THz	nm	THz	nm
1	186.00	1611.79	186.05	1611.35	191.00	1569.59	191.05	1569.18
2	186.10	1610.92	186.15	1610.49	191.10	1568.77	191.15	1568.36
3	186.20	1610.06	186.25	1609.62	191.20	1576.95	191.25	1567.54
4	186.30	1609.19	186.35	1608.76	191.30	1567.13	191.35	1566.72
5	186.40	1608.33	186.45	1607.90	191.40	1566.31	191.45	1565.90
6	186.50	1607.47	186.55	1607.04	191.50	1565.50	191.55	1565.09
7	186.60	1606.60	186.65	1606.17	191.60	1564.68	191.65	1564.27
8	186.70	1605.74	186.75	1605.31	191.70	1563.86	191.75	1563.45
9	186.80	1604.88	186.85	1604.46	191.80	1563.05	191.85	1562.64
10	186.90	1604.03	186.95	1603.60	191.90	1562.23	191.95	1561.83

3.2.1 Dense WDM

The term *dense WDM* (DWDM) refers to the close optical frequency spacings denoted by ITU-T Recommendation G.694.1, which is aimed specifically at DWDM. This document specifies WDM operation in the S-, C-, and L-bands for high-quality, high-rate metro area network (MAN) and wide area network (WAN) services. It lists specifications for narrow frequency spacings of 100 to 12.5 GHz (or, equivalently, 0.8 to 0.1 nm at 1550 nm). This implementation requires the use of stable, high-quality, temperature- and wavelength-controlled (frequency-locked) laser diode light sources. For example, the wavelength-drift tolerances for 25-GHz channels are ±0.02 nm.

Table 3.1 lists part of the by ITU-T G.694.1 DWDM frequency grid for 100- and 50-GHz spacings in the L- and C-bands. The column labeled "50-GHz offset" means that for the 50-GHz grid one uses the 100-GHz spacings with these 50-GHz values interleaved. For example, the 50-GHz channels in the L-band would be at 186.00, 186.05, 186.10 THz, and so on. Note that when the frequency spacings are uniform, the wavelengths are not spaced uniformly because of the relationship given in Eq. (3.1).

To designate which channel is under consideration in 100-GHz applications, the ITU-T uses a *channel numbering convention*. For this, the frequency 19N.M THz is designated as ITU channel number NM. For example, the frequency 194.3 THz is ITU channel 43.

3.2.2 Coarse WDM

With the production of full-spectrum (low-water-content) G.652C and G.652D fibers, the development of relatively inexpensive optical sources, and the desire to have low-cost optical links operating in access networks and local area networks, came the concept of *coarse WDM* (CWDM). In 2002 the ITU-T released

STANDARD WDM SPECTRAL GRIDS

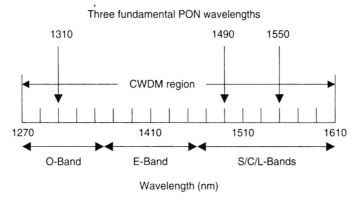

Figure 3.3. Spectral grid for CWDM.

Recommendation G.694.2 which defines the spectral grid for CWDM. As shown in Figure 3.3, the CWDM grid is made up of 18 wavelengths defined within the range 1270 to 1610 nm (O- through L-bands) spaced by 20 nm with wavelength-drift tolerances of ±2 nm. This can be achieved with inexpensive light sources that are not temperature controlled.

The ITU-T Recommendation G.695 released in 2003 outlines optical interface specifications for multichannel CWDM over distances of 40 and 80 km. Both unidirectional and bidirectional systems (such as used in PON applications) are included in the recommendation. The applications for G.695 cover all or part of the 1270- to 1610-nm wavelength range. The main deployments are for single-mode fibers, such as those specified in ITU-T Recommendations G.652 and G.655.

3.2.3 PON Spectral Regions

Figure 3.3 also indicates the values of the three fundamental wavelengths used for PON applications. These are 1490 and 1550 nm for downstream (central office to user) transmission and 1310 nm for the upstream (user to central office) link. Typically, the transceivers for the sending information downstream are located within a central office, which has a fairly stable temperature environment.

In contrast, the 1310-nm transmitter often is within an interface box that is located outdoors. This means that such a device can experience a large temperature variation of tens of degrees Celsius. Since a laser that is not temperature controlled typically exhibits a wavelength drift of 0.1 nm/°C, the wavelength can vary significantly from its nominal 1310-nm operational point. However, the 1310-nm point is far enough away from the 1490-nm source that the wavelength drift is not a problem. It also should be noted that although a PON typically does not use a thermoelectric cooler to maintain the laser diode at a fixed temperature for wavelength stabilization, a transmitter can contain circuitry for controlling the modulation and bias currents of the laser diode to guarantee a constant output power of the laser over temperature fluctuations and aging effects.

3.3 OPTICAL COUPLERS

An *optical coupler* is a key device for WDM networks. This device category encompasses a variety of functions, including splitting a light signal into two or more streams, combining two or more light streams, tapping off a small portion of optical power for monitoring purposes, or transferring a selective range of optical power from one fiber to another.

When discussing couplers it is customary to refer to them in terms of the number of input ports and output ports on the device. For example, a device with two inputs and two outputs would be called a 2×2 *coupler*. In general, an $N \times M$ coupler has $N \geq 1$ input ports and $M \geq 2$ output ports. The coupling devices can be fabricated either from optical fibers or by means of planar optical waveguides using material such as lithium niobate ($LiNbO_3$) or indium phosphide (InP). A planar optical waveguide device often is referred to as a *planar lightwave circuit* (PLC). The couplers described in this section include 2×2 couplers and tap couplers. In Chapter 5 we give operational and application details of specific couplers for PONs, such as star couplers and three-wavelength multiplexing devices.

3.3.1 Basic 2 × 2 Coupler

The 2×2 coupler is a simple fundamental device that we use here to demonstrate the operational principles of optical couplers. These also are known as *directional couplers*. A common construction is the *fused-fiber coupler* illustrated in Figure 3.4. This device is fabricated by twisting together, melting, and pulling two single-mode fibers so they get fused together over a uniform section of length W. Each input and output fiber has a long tapered section of length L, since the transverse dimensions are reduced gradually down to that of the coupling region when the fibers are pulled during the fusion process. As shown in Figure 3.4, P_0 is the input power on the top fiber (which we will take as the primary fiber in a link), P_1 is the throughput power, and P_2 is the power coupled into the second fiber. The parameters P_3 and P_4 are

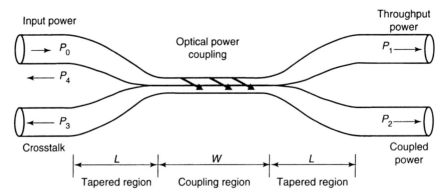

Figure 3.4. Cross-sectional view of a fused-fiber coupler and its operation concept.

extremely low optical signal levels (-50 to -70 dB or, equivalently, factors of 10^{-5} to 10^{-7} below the input power level). These result from backward reflections and scattering due to packaging effects and bending in the device.

To understand how power is coupled from one fiber to the other, recall the diagram of the optical power distribution pattern shown in Figure 2.12. This figure shows that the power distributions of any given mode are not confined completely to the fiber core, but instead, they extend partially into the cladding. Therefore, if two fiber cores are brought close together, the tail of the power distribution in one fiber will extend into the adjacent fiber core. Consequently, some of the optical power will transfer to the adjacent fiber through *evanescent coupling*. The amount of optical power coupled from one fiber to another can be varied by changing the coupling length W or the distance between the two fiber cores (see Problem 3.4).

3.3.2 Coupler Performance

In specifying the performance of an optical coupler, one usually indicates the percentage division of optical power between the output ports by means of the *splitting ratio* or *coupling ratio*. Referring to Figure 3.4, with P_0 being the input power and P_1 and P_2 the output powers, the

$$\text{coupling ratio} = \frac{P_2}{P_1 + P_2} \times 100\% \tag{3.2}$$

By adjusting the parameters so that power is divided evenly, with half of the input power going to each output, one creates a *3-dB coupler*. A coupler also could be made in which almost all the optical power at 1550 nm (for example) goes to one port and almost all the energy around 1310 nm goes to the other port.

In the analysis above, we have assumed for simplicity that the device has no internal loss. However, in any practical coupler there is always some light that is lost when a signal goes through it. The two basic losses are excess loss and insertion loss. The *excess loss* is defined as the ratio of the input power to the total output power. Thus, in decibels, the excess loss for a 2×2 coupler is

$$\text{excess loss} = 10 \log \frac{P_0}{P_1 + P_2} \tag{3.3}$$

The *insertion loss* refers to the loss for a particular port-to-port path. For example, for the path from input port i to output port j, we have, in decibels,

$$\text{insertion loss} = 10 \log \frac{P_i}{P_j} \tag{3.4}$$

Another performance parameter is *crosstalk*, which measures the degree of isolation between the input at one port and the optical power scattered or reflected back into the other input port. That is, it is a measure of the optical power level P_3 shown in Figure 3.4:

$$\text{crosstalk} = 10 \log \frac{P_3}{P_0} \tag{3.5}$$

As an example, suppose that a 2 × 2 fused-fiber coupler has an input optical power level of $P_0 = 200\,\mu\text{W}$. The output powers at the other three ports are $P_1 = 90\,\mu\text{W}$, $P_2 = 85\,\mu\text{W}$, and $P_3 = 6.3\,\text{nW}$. From Eq. (3.2), the coupling ratio is

$$\text{coupling ratio} = \frac{85}{90 + 85} \times 100\% = 48.6\%$$

From Eq. (3.3), the excess loss is

$$\text{excess loss} = 10 \log \frac{200}{90 + 85} = 0.5\,\text{dB}$$

Using Eq. (3.4), the insertion losses are

$$\text{insertion loss (port 0 to port 1)} = 10 \log \frac{200}{90} = 3.47\,\text{dB}$$
$$\text{insertion loss (port 0 to port 2)} = 10 \log \frac{200}{85} = 3.72\,\text{dB}$$

The crosstalk is given by Eq. (3.5) as

$$\text{crosstalk} = 10 \log \frac{6.3 \times 10^{-3}}{200} = -45\,\text{dB}$$

3.3.3 Tap Coupler

To monitor the light signal level or quality in a link, one can use a 2 × 2 device that has a coupling fraction of around 1 to 5 percent, which is selected and fixed during fabrication. This is known as a *tap coupler*. Nominally, the tap coupler is packaged as a three-port device with one arm of the 2 × 2 coupler being terminated inside the package. Figure 3.5 shows a typical package for such a tap coupler, and Table 3.2 lists some representative specifications.

An alternative is to use a tap coupler that has an integrated micro-tap and a *pin* photodiode in a single package. Figure 3.6 illustrates such a device, which has a 3.5-mm diameter and a length of 18 mm. Such a compact device reduces the number of parts that need to be handled, can easily be spliced into the fiber line, improves fiber routing, and saves board space on a circuit card. Table 3.3 lists some performance characteristics of such a device, which can be used for C-band, L-band, and C+L-band applications. Devices nominally operating at 5 V and having

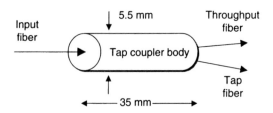

Figure 3.5. Typical package for a simple tap coupler.

TABLE 3.2 Representative Specifications for a 2 × 2 Tap Coupler

Parameter	Unit	Specification
Tap ratio	%	1 to 5
Insertion loss	dB	0.5
Return loss	dB	55
Power handling	mW	1000
Flylead length	m	1
Size (diameter × length)	mm	5.5 × 35

TABLE 3.3 Generic Performance Characteristics of a Compact Integrated Tap Coupler

Tap Ratio (%)	Maximum Insertion Loss (dB)				Maximum Responsivity (nA/mW)	Maximum Optical Input (mW)
	C- or L-Band		C+L-Band			
1	0.3	0.4	0.4	0.5	0.8	500
3	0.4	0.5	0.5	0.6	24	300
5	0.5	0.6	0.6	0.7	45	300

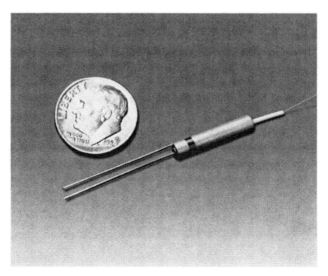

Figure 3.6. Compact tap coupler consisting of a microtap and a pin photodiode. (Photo courtesy of Lightwaves2020; www.lightwaves2020.com.)

a wavelength-dependent loss of less than 0.15 dB can be made with tap ratios ranging from 1 to 5 percent.

3.4 BIDIRECTIONAL WDM LINKS

The use of different wavelengths for sending independent streams of information also allows full-duplex (bidirectional) communication over a single fiber. Such an

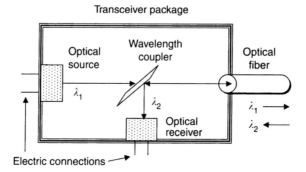

Figure 3.7. Concept of a bidirectional link implementation.

implementation will result in an overall system cost reduction, since it eliminates one fiber from a standard two-fiber full-duplex link. Another advantage is that all the optical components can be integrated into a single package, thereby reducing costs. Thus, a transmission link provider can double the capacity of a link without installing new fibers. This is the driving force behind using single-fiber bidirectional WDM links for PON applications.

Figure 3.7 illustrates the *bidirectional link* scheme. Here a transceiver module contains both an optical source and a photodetector. Light at wavelength λ_1 is inserted into a fiber and optical signals at wavelength λ_2 are received. In this simple case, the wavelength coupler is highly transparent to wavelength λ_1 and highly reflective for wavelength λ_2.

The factor that must be kept in mind with this bidirectional scheme, however, is that the wavelength coupling elements at either end of the link will reduce the available optical power budget, since they have excess and insertion losses. In Chapter 11 we give more details on the fiber-to-the-premises (FTTP) link design analyses.

3.5 SUMMARY

An attractive aspect of an optical communication link is that wavelength-division multiplexing (WDM) allows many different independent optical information channels to be sent along a single fiber simultaneously. This is done by selecting different wavelengths from the spectral regions ranging from the O-band through the L-band.

A major design criterion of a WDM system is that the wavelengths must be spaced properly to avoid interference between adjacent channels. A basic PON design accommodates this through the selection of wavelengths at 1310, 1490, and 1550 nm. These large spacings allow the peak operating wavelengths of the end components to drift widely with factors such as large temperature changes and manufacturing-related component variations. This enables the designers to use inexpensive transceivers that do not require stringent wavelength control.

Passive optical networks use the concept of coarse WDM (CWDM). The ITU-T Recommendation G.694.2 for CWDM defines the spectral grid for CWDM. This is made up of 18 wavelengths within the range 1270 to 1610 nm (O- through L-bands) spaced by 20 nm with wavelength-drift tolerances of ±2 nm. The ITU-T Recommendation G.695 outlines optical interface specifications for multichannel CWDM over distances of 40 and 80 km. Both unidirectional and bidirectional systems (such as used in PON applications) are included in the recommendation. The applications for G.695 cover all or part of the wavelength range 1270 to 1610 nm. The main deployments are for single-mode fibers such as those specified in ITU-T Recommendations G.652 and G.655.

PROBLEMS

3.1 A DWDM optical transmission systems is designed to have 100-GHz channel spacings. Using Eq. (3.1), find how many wavelength channels can be utilized in the 1536- to 1556-nm spectral band.

3.2 Assume that a 32-channel DWDM system has uniform channel spacings of $\Delta v = 100$ GHz, and let the frequency v_n correspond to the wavelength λ_n. Using this correspondence, let the wavelength $\lambda_1 = 1550$ nm. Calculate the wavelength spacing between the first two channels (channels 1 and 2) and between the last two channels (channels 31 and 32). From the result, what can be concluded about using an equal-wavelength spacing definition in this wavelength band instead of the standard equal-frequency channel spacing specification?

3.3 Consider the fused-fiber coupler shown in Figure 3.4. Assuming that the coupler is lossless, the power P_2 coupled into the second fiber over a length x can be found from the expression $P_2 = P_0 \sin^2 kx$, where k is the coupling coefficient describing the interaction between the electromagnetic fields in the two fibers.

(a) Since the coupler is lossless, show that the throughput power is $P_1 = P_0 \cos^2 kx$. What does this indicate about the phases of P_1 and P_2?

(b) Suppose that for a 1310-nm wavelength we have $k = \pi/24.8$ mm. Plot the expressions for P_1 and P_2 for coupling lengths $0 \leq x \leq 15$ mm.

3.4 Consider the coupling ratios as a function of pull lengths shown in Figure 3.8 for a fused-fiber coupler. The performances are given for 1310- and 1540-nm operation. Discuss the behavior of the coupler for each wavelength if its pull length is stopped at the following points: A, B, C, D, E, and F.

3.5 The data sheet for a 2 × 2 single-mode fused-fiber coupler with a 40:60 splitting ratio states that the insertion losses are 2.7 dB for the 60 percent channel and 4.7 dB for the 40 percent channel.

(a) If the input power equals $P_0 = 200\,\mu$W, find the output levels P_1 and P_2.

(b) Find the excess loss of the coupler.

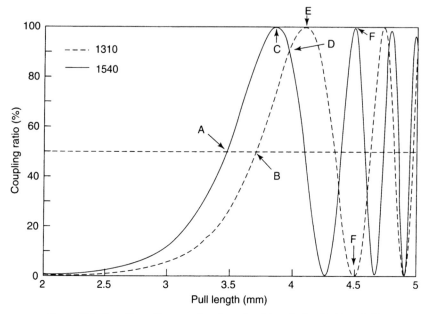

Figure 3.8. Coupling ratios as a function of pull length (for Problem 3.4).

(c) From the calculated values of P_1 and P_2, verify that the splitting ratio is 40:60.

3.6 A laser transmitter that is not temperature controlled has a wavelength drift equal to 0.1 nm/°C. Suppose that this device is used in an outdoor PON interface box that can experience a temperature variation from -20 to $+30°C$. If the wavelength is 1310 nm at 20°C, what is the wavelength change over this temperature range?

3.7 Assume that for a given tap coupler the throughput and coupled powers are 230 and 5 µW, respectively, for an input power of 250 µW.
  (a) What is the coupling ratio?
  (b) What are the insertion losses?
  (c) Find the excess loss of the coupler.

3.8 Using a Web search tool, find some vendors of fused-fiber couplers and from their data sheets list some coupler performance characteristics. Consider parameters such as number of coupling branches, coupling ratio, insertion loss, return loss, package configuration, and physical size.

3.9 Using a Web search tool, find some vendors of tap couplers and from their data sheets list representative performance specifications for several different types. Consider characteristics such as tap ratio (use 1 to 5 percent), insertion loss, return loss, optical power handing limit, package configuration, and physical size.

3.10 Planar lightwave circuit (PLC) technology can realize manufacturing cost savings by integrating several optical functions (such as filtering, amplification, coupling, and attenuation) on a single substrate. Using Web resources, examine the physical characteristics of PLC technology, for example, what are PLC devices made of, and how are they fabricated. List the operational characteristics of some commercially available devices, in particular, look at PLC-based optical coupling and power splitter devices.

FURTHER READING

1. ITU-T Recommendation G.692, "Optical Interfaces for Multichannel Systems with Optical Amplifiers," Mar. 1997.
2. ITU-T Recommendation G.694.1, "Dense Wavelength Division Multiplexing (DWDM)," 2002.
3. ITU-T Recommendation G.694.2, "Coarse Wavelength Division Multiplexing (CWDM)," 2002.
4. ITU-T Recommendation G.695, "Optical Interfaces for Coarse Wavelength Division Multiplexing Applications," Feb. 2004.
5. G. E. Keiser, "A review of WDM technology and applications," *Opt. Fiber Technol.*, vol. 5, pp. 3–39, Jan. 1999.
6. O. Krauss, *DWDM and Optical Networks*, Wiley, Hoboken, NJ, 2002.
7. J. Zheng and M. T. Mouftah, *Optical WDM Networks: Concepts and Design Principles*, Wiley, Hoboken, NJ, 2004.

4

PON TRANSCEIVERS

A basic passive optical network uses three separate wavelengths on a single fiber. From the central office an optical source operating at 1490 nm sends combined voice and data traffic to users and a 1550-nm source transmits video content to subscribers. The return link from the users to the central office employs a 1310-nm wavelength for combined voice and data traffic. Thus, at the central office end of a link there are two optical sources and one receiver, whereas the customer equipment contains two optical receivers and one light source. Recall that to distinguish the traffic-flow directions, the term *downstream* refers to transmissions from the central office to the users and the term *upstream* is the transmission path from the users to the central office. As we describe in detail in later chapters, the transmission equipment in a passive optical network consists of an *optical line terminal* (OLT) situated at the central office and an *optical network terminal* (ONT) at the customer premises.

Typically, in the central office the two optical sources are on separate circuit cards that reside in different equipment racks. To reduce circuit board size and since there is only one bidirectional fiber, optical transmitters and receivers in the ONT equipment at the customer's premises often are combined into a single transceiver package. Included in such a package would be optical source modules, photodetection modules, a passive device for combining and separating the individual wavelengths, digital clock and data recovery interfaces for data and voice, alarm detection and laser shutdown circuitry, and video receiver circuitry.

In this chapter, in Section 4.1 we address the requirements and operating characteristics of optical sources for PONs. The topics of Section 4.2 deal with the performance issues of photodetectors and optical receivers. A figure of merit for receivers is the bit error rate. In Section 4.3 we define this parameter in terms of the optical signal-to-noise ratio. For PON applications, a specially designed burst-mode receiver is needed at the OLT, since the amplitude and phase of information packets

---

FTTX Concepts and Applications, by Gerd Keiser
Copyright © 2006 John Wiley & Sons, Inc.

arriving from different user locations can vary widely from packet to packet. This is the topic of Section 4.4. In contrast to conventional transmitters, in a PON the lasers at an ONT must be turned off almost completely when they are not transmitting. The basis of this burst-mode transmitter function is the topic of Section 4.5. Since a PON is based on the use of bidirectional optical fiber links, in Section 4.6 we address special transceiver configurations that may include the light source, the optical receiver functions, and the wavelength multiplexer in a single package.

4.1 OPTICAL SOURCES FOR PONS

Light sources used in optical communications can range from simple inexpensive light-emitting diodes (LEDs) to costly high-power laser diodes with complex semiconductor structures. The corresponding transmitters using such sources vary from a simple inexpensive LED-based package for short-distance links to expensive laser diode-based modules that contain sophisticated electronics for controlling the temperature, wavelength stability, and optical output power level for long-distance DWDM links. For PON applications the transmitters fall in between these two extremes, with a major emphasis being low cost.

4.1.1 Source Characteristics

Semiconductor-based *laser diodes* are used widely as light sources in optical fiber communication systems. The main laser types applicable to PONs are the Fabry–Perot (FP) laser and the distributed-feedback (DFB) laser. Key properties of these lasers include high optical output powers, narrow linewidths, and highly directional output beams for efficient coupling of light into fibers.

The wide wavelength spacing in PON applications results in the following performance and cost issues for lasers in contrast to more expensive long-haul and metro links:

1. DFB or FP lasers that are not temperature-controlled can be used since it is not essential to maintain the laser output at a precise wavelength.
2. Since a laser-cooling module is not needed for stringent laser stabilization, nominal electric power requirements are 0.5 W compared to 4 W for temperature-controlled lasers.
3. Whereas wavelength tolerances of lasers for DWDM use are on the order of ± 0.1 nm, the manufacturing tolerances for PON lasers are around ± 3 nm. This reduces yield costs significantly.
4. The elimination of laser-cooling requirements and the higher laser manufacturing yields reduce both the cost and size of PON transmitters by a factor of 4 or 5 compared to DWDM devices.
5. Wideband optical filter technology can be used for wavelength multiplexing at about half the cost of DWDM components.

4.1.2 DFB and FP Lasers

To understand the differences between the two basic lasers used for PONs, we need to look at the lasing process in these devices. The word *laser* is an abbreviation for *light amplification by stimulated emission of radiation*. So what does this mean? In a semiconductor-based device the laser action takes place within a region called the *gain medium* or *laser cavity*. To achieve lasing, the photon density in this region needs to be built up so that the stimulated emission rate becomes higher than the rate at which photons are absorbed by the semiconductor material. *Stimulated emission* occurs when some external stimulant (such as an incident photon) causes an electron located in an excited state to drop to the ground state. This process results in the emission of a photon that has an energy equal to the difference between the excited state and the ground state electron energies. For stimulated emission to occur, there must be a *population inversion* of carriers. This simply means that there are more electrons in an excited state than in the ground state. Since this is not a normal condition, it is achieved by supplying additional external energy to raise electrons to a higher energy level. This energy-supplying technique, called *pumping*, can be done by optical or electrical means. For example, the bias voltage from a power supply provides the externally pumped energy in a semiconductor laser.

A variety of mechanisms can be used either at the ends or within the laser cavity to establish a population inversion. The purpose of this mechanism is to reflect most of the emitted photons back and forth through the gain medium. With each pass through the cavity the photons stimulate more excited electrons to drop to the ground state, thereby emitting more photons of the same wavelength. This process thus builds up the photon density in the gain region until lasing occurs.

In the *Fabry–Perot laser* the lasing cavity is defined by the two end faces of a miniature semiconductor chip such as InGaAsP, as shown in Figure 4.1. These end faces, called *facets*, act as light-reflecting mirrors. This structure is called a *Fabry–Perot cavity* or an *etalon*. Since this cavity is fairly long (around 250 µm), the laser will oscillate (or *resonate*) simultaneously in several modes or frequencies. This

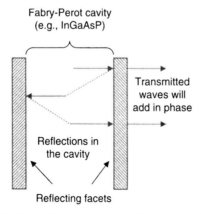

Figure 4.1. Two parallel light-reflecting surfaces define a Fabry–Perot cavity or etalon.

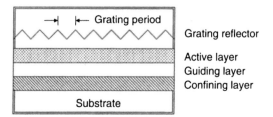

Figure 4.2. Series of closely spaced reflectors provides light feedback for lasing in a DFB laser.

effect produces a broad Gaussian-shaped output spectrum that has a width of several nanometers at its half-maximum point. Although this broad spectral output does not make an FP laser feasible for high-speed long-haul transmissions, this device is popular for use at short and intermediate distances (up to 20 km). In PONs an FP laser may be used for 1310-nm upstream and 1490-nm downstream digital links running at data rates of up to 1.25 Gbps. These applications use the small-form-factor transceiver package described in Section 4.4. Optical output powers from such a transceiver can range from -10 to -3 dBm (0.1 to 0.5 mW). A typical linewidth of an FP laser at its half-maximum power level is <4 nm at a 1310-nm emission wavelength and <8 nm at an emission wavelength of 1490 nm.

In a *distributed-feedback laser*, a series of closely spaced reflectors provides light feedback in a distributed fashion throughout the semiconductor cavity for enabling lasing, as shown in Figure 4.2. Through a suitable design of these reflectors, the device can be made to oscillate in only a single mode with a very narrow linewidth. This means that it emits at a fairly well-defined wavelength. The particular operating wavelength can be selected at the time of device fabrication by an appropriate choice of the reflector spacing. Single-mode DFB lasers are used extensively in high-speed transmission systems and for sending video signals at 1550 nm in a PON. Depending on the packaging format, DFB outputs can range up to several milliwatts: for example, 2 mW (3 dBm). The DFB linewidth usually is $\ll 1$ nm.

4.1.3 Modulation Speed

For PON applications, the light source operates in a *direct modulation* mode, which is the process of applying a varying electrical signal to the laser to change the optical output level. The term *modulation speed* refers to how fast a device can be turned on and off by an electrical signal to produce a corresponding optical output pattern. Laser diodes can be modulated significantly faster than LEDs. However, there is a speed limit beyond which even the laser does not respond fast enough to the changes in an electrical signal. This speed limit is about 2.5 Gbps. Beyond this point a steady light output stream from the laser diode is fed into an external device, which can very rapidly change the intensity of the light that passes through it. This process is known as *external modulation*.

To have PON transceivers be inexpensive, the data rates were chosen so that more complex and costly external modulators are not required. As discussed in later

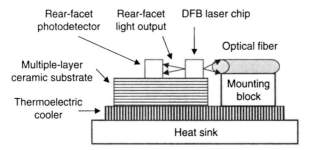

Figure 4.3. High-performance optical transmitter containing a DFB laser, a fiber mounting block, a thermoelectric cooler, and a monitoring photodiode.

chapters, the maximum data rates considered for PONs is 2.5 Gbps. Therefore, in almost all cases a direct modulation mode can be used in the transmitter.

4.1.4 Optical Transmitter Packages

Normally, an equipment manufacturer does not obtain an isolated light source, but purchases it as part of a standard optical transmitter package. This package provides the following: a mounting block for the semiconductor light source chip, a holder for attaching a light-coupling fiber, and various control electronics. In some cases the transmitter package also contains a miniature thermoelectric cooler (TEC) for maintaining the temperature at a fixed value. These components typically are mounted on a metallic platform to create a laser module. The module also may contain a second photodiode for precise monitoring of the output power level. Figure 4.3 shows a high-performance module containing a DFB laser, a fiber mounting block, a thermoelectric cooler, and a monitoring photodiode. For PON applications these modules are mounted inside the various transceiver packages described in Section 4.5.

4.2 OPTICAL RECEIVERS

The first element of the optical receiver is a photodetector. The *photodetector* senses the light signal falling on it and converts the variation of the arriving optical power into a correspondingly varying electric current. Since the optical signal generally is weakened and distorted when it emerges from the end of the fiber, the photodetector must meet strict performance requirements. Among the most important of these are:

- A high sensitivity to the emission wavelength range of the light signal received, to provide an efficient photon-to-electron conversion process
- A minimum addition of noise to the signal by the photodetector and the following signal-processing electronics
- A fast response speed to handle the data rate desired, to avoid undesirable distortions in the electrical output from the photodetector

4.2.1 Photodetector Types

Semiconductor-based photodiodes are the main devices that satisfy the set of requirements listed above. The most common one is the *pin photodiode*, since it is a stable and highly reliable device. A second semiconductor photodetector class is the *avalanche photodiode* (APD), which internally multiplies the primary signal photocurrent before it enters the input circuitry of the following amplifier. This *avalanche multiplication* effect increases receiver sensitivity since the photocurrent is multiplied prior to encountering the electrical noise associated with the receiver circuitry. The multiplication factor or *gain M* can be made quite large, but larger gains increase the noise currents of the device because of larger variations in the amplified photocurrent. Thus, an APD has a *noise figure* F(M) that is associated with the random nature of the avalanche process.

The choice of a photodetector material is important since its semiconductor properties determine the wavelength range over which the device will operate. For operation in the region 1100 to 1700 nm the ternary material indium gallium arsenide (InGaAs) is used most commonly for both pin and avalanche photodiodes. Tables 4.1 and 4.2 list some of the operating parameters of InGaAs pin and avalanche photodiodes, respectively. The values were derived from various vendor data sheets and from performance numbers reported in the literature. They are given as guidelines for comparison purposes. Detailed values on specific devices can be obtained from photodetector and receiver module suppliers.

TABLE 4.1 Generic Operating Parameters of an InGaAs pin Photodiode

Parameter	Symbol	Unit	Value Range
Wavelength range	λ	nm	1100–1700
Responsivity	$\mathcal{R}$	A/W	0.75–0.95
Dark current	I_D	nA	0.5–2.0
Rise time	τ_r	ns	0.05–0.5
Bandwidth	B	GHz	1–2
Bias voltage	V_B	V	5

TABLE 4.2 Generic Operating Parameters of an InGaAs Avalanche Photodiode

Parameter	Symbol	Unit	Value Range
Wavelength range	λ	nm	1100–1700
Avalanche gain	M	—	10–40
Dark current	I_D	nA	10–50 at $M=10$
Rise time	τ_r	ns	0.1–0.5
Gain–bandwidth	MB	GHz	20–250
Bias voltage	V_B	V	20–30

4.2.2 Quantum Efficiency

An important characteristic of a photodetector is its *quantum efficiency*, which is designated by the symbol η. The quantum efficiency is the number of electron–hole (eh) pairs that are generated per incident photon of energy hc/λ and is given by

$$\eta = \frac{\text{number of eh pairs generated}}{\text{number of incident photons}} = \frac{I_p/q}{P_0/(hc/\lambda)} \tag{4.1}$$

Here q is the electron charge, I_p is the average photocurrent generated by a steady photon stream of average optical power P_0 incident on the photodetector, $h = 6.6256 \times 10^{-34}$ J·s is Planck's constant. In a practical photodiode, 100 photons will create between 30 and 95 electron–hole pairs. This gives a quantum efficiency ranging from 30 to 95 percent.

4.2.3 Responsivity

The performance of a photodiode may be characterized by its *responsivity* $\mathcal{R}$. This is related to the quantum efficiency by

$$\mathcal{R} = \frac{I_p}{P_0} = \frac{\eta q}{hc/\lambda} \tag{4.2}$$

This parameter is quite useful since it specifies the photocurrent generated per unit optical power. Figure 4.4 shows typical pin photodiode responsivities as a function of wavelength. Representative values for InGaAs are 0.9 A/W at 1300 nm and about 0.95 A/W at 1490 and 1550 nm.

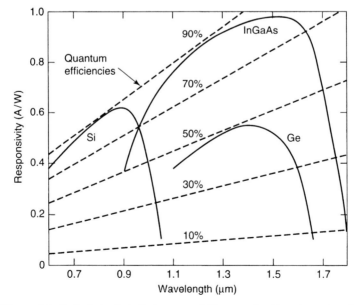

Figure 4.4. Typical pin photodiode responsivities as a function of wavelength.

In most photodiodes the quantum efficiency is independent of the power level falling on the detector at a given photon energy. Thus, the responsivity is a linear function of the optical power. That is, the photocurrent I_p is directly proportional to the optical power P_0 incident on the photodetector. This means that the responsivity $\mathcal{R}$ is constant at a given wavelength. Note, however, that the quantum efficiency is not a constant at all wavelengths, since it varies according to the photon energy. Consequently, the responsivity is a function of the wavelength and of the photodiode material. For a given material, as the wavelength of the incident photon becomes longer, the photon energy becomes less than that required to excite an electron from the valence band to the conduction band. The responsivity thus falls off rapidly beyond the cutoff wavelength, as can be seen in Figure 4.4.

Analogous to the pin photodiode, the performance of an APD is characterized by its responsivity $\mathcal{R}_{ADP}$. In an APD the multiplied photocurrent I_M is given by

$$I_M = \mathcal{R}_{ADP} P_0 = M\mathcal{R} P_0 \tag{4.3}$$

where $\mathcal{R}$ is the unity-gain responsivity.

4.2.4 Speed of Detector Response

To interpret high-rate signals properly, photodiodes need to have a fast response speed. For example, for PON applications the data rates can be as high as 2.5 Gbps. If the detector output does not track closely the variations of the incoming optical pulse shape, the shape of the output pulse will be distorted. This will affect the link performance since it may introduce errors when the receiver electronics interpret the optical signal. The detector response speed is measured in terms of the time it takes the output signal to rise from 10 percent to 90 percent of its peak value when the input to the photodiode is turned on instantaneously. This is shown in Figure 4.5 and is known as the *10 to 90 percent rise time*. Similarly, the time it takes the output to drop from its 90 percent value to its 10 percent value is known as the *fall time*.

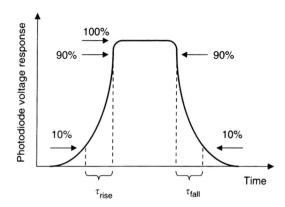

Figure 4.5. Photodetector response to an optical pulse showing the 10 to 90 percent rise time and the 90 to 10 percent fall time.

The rise and fall times depend on factors such as how much of the light is absorbed at a specific wavelength, the width of the intrinsic photon-absorption region, photodiode and electronics capacitance values, and detector and electronics resistances. As a result, the rise and fall times are not necessarily equal in a receiver. For example, large capacitance values can cause a long decay tail to appear in the falling edge of the output pulse, thereby creating long fall times.

4.2.5 Receiver Bandwidth

The response speed limitations of the photodiode and the electronic components result in a gradual drop in the output level beyond a certain frequency. The point at which the output has dropped to 50 percent of its low-frequency value is called the *3-dB point*. At this point only half as much signal power is getting through the detector compared to lower frequencies. The 3-dB point defines the receiver *bandwidth* (sometimes referred to as the *3-dB bandwidth*), which defines the range of frequencies that a receiver can reproduce in a signal. If the rise and fall times are equal, the 3-dB bandwidth (in MHz) can be estimated from the rise time by the relationship

$$\text{bandwidth (MHz)} = \frac{350}{\text{rise time (ns)}} \quad (4.4)$$

where the rise time is expressed in units of nanoseconds.

4.2.6 Photodetector Noise

In fiber optic communication systems, the photodiode must detect very weak optical signals. In the process of converting the received optical signal power into an electrical signal, various noises and distortions will unavoidably be introduced, due to imperfect component responses. This can lead to errors in the interpretation of the signal that is received. The term *noise* describes unwanted components of a signal that tend to disturb the transmission and processing of the signal in a physical system. Noise is present in every communication system and represents a basic limitation on the transmission and detection of signals. In electric circuits noise is caused by the spontaneous fluctuations of current or voltage.

Detection of the weakest possible optical signals requires that the photodetector and its associated electronic amplification circuitry be optimized so that a specific signal-to-noise ratio is maintained. The *signal-to-noise ratio* (SNR) at the output of an optical receiver is defined by

$$\text{SNR} = \frac{\text{signal power from photocurrent}}{\text{photodetector noise power} + \text{amplifier noise power}} \quad (4.5)$$

The noises in the receiver arise from the statistical nature of the randomness of the photon-to-electron conversion process and the electronic noise in the receiver amplification circuitry.

To achieve a high SNR, the numerator in Eq. (4.5) should be maximized and the denominator should be minimized. Thus, the following conditions should be met:

1. The photodetector must have a high quantum efficiency to generate a large signal power.
2. The photodetector and amplifier noises should be kept as low as possible.

For most applications it is the noise currents in the receiver electronics that determine the minimum optical power that can be detected, since the photodiode quantum efficiency normally is close to its maximum possible value.

The main noises associated with pin photodiode receivers are quantum or shot noise and dark current associated with photodetection, and thermal noise occurring in the electronics. *Shot noise* arises from the statistical nature of the production and collection of photoelectrons. It has been found that these statistics follow a Poisson process. Since the fluctuation in the number of photocarriers created is a fundamental property of the photodetection process, it sets the lower limit on the receiver sensitivity when all other conditions are optimized.

The photodiode *dark current* arises from electrons and holes that are thermally generated at the pn junction of the photodiode. This current continues to flow through the bias circuit of the device when no light is incident on the photodetector. In an APD these liberated carriers also get accelerated by the electric field across the device and therefore are multiplied by the avalanche mechanism. *Thermal noise* arises from the random motion of electrons that is always present at any finite temperature.

4.3 RECEIVER BER AND OSNR

The most common figure of merit for digital links is the *bit error rate* (BER). This is defined as the number of bit errors N_E occurring over a specific time interval divided by the total number of bits N_T sent during that interval, that is, BER = N_E/N_T. The error rate is expressed by a number, such as 10^{-9}, which states that on the average one error occurs for every billion pulses sent. Typical error rates specified for optical fiber telecommunication systems range from 10^{-9} to 10^{-15}.

An important point about the BER is that it is determined principally by the *optical signal-to-noise ratio* (OSNR). Therefore, it is the OSNR that is measured when a WDM link is installed and when it is in operation. The OSNR does not depend on factors such as the data format, pulse shape, or optical filter bandwidth, but only on the average optical signal power P_{signal} and the average optical noise power P_{noise}. The TIA/EIA-526-19 standard defines OSNR over a given reference spectral bandwidth B_{ref} (normally, 0.1 nm) as (in decibels)

$$\text{OSNR(dB)} = 10 \log \frac{P_{signal}}{P_{noise}} + 10 \log \frac{B_m}{B_{ref}} \qquad (4.6)$$

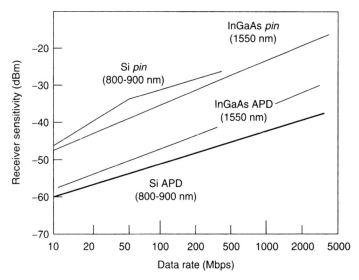

Figure 4.6. Receiver sensitivities at several different wavelengths of silicon and InGaAs pin and avalanche photodiodes as a function of bit rate. The Si pin, Si APD, and InGaAs pin curves are for a 10^{-9} BER. The InGaAs APD curve is for a 10^{-11} BER.

where B_m is the noise-equivalent measurement bandwidth of the instrument being used: for example, an optical spectrum analyzer. The noise may be from sources such as the transmitter, crosstalk, or amplified spontaneous emission (ASE) from an erbium-doped fiber amplifier (EDFA). OSNR is a metric that can be used in the design and installation of networks as well as to check the health and status of individual optical channels.

The sensitivity of a photodetector in an optical fiber communication system is describable in terms of the *minimum detectable optical power*. This is the optical power necessary to produce a photocurrent equal to the total noise current, or equivalently, to yield a SNR of 1. Figure 4.6 shows the sensitivities of silicon and InGaAs pin and avalanche photodiodes at several different wavelengths. All curves are for bit error rates of 10^{-9} except for the InGaAs APD curve, which is for a 10^{-11} BER.

4.4 BURST-MODE RECEIVER CONCEPT

For PON applications, the operational characteristics of an OLT optical receiver differ significantly from those used in conventional point-to-point links. This arises from the fact that the amplitude and phase of information packets received in successive time slots from different user locations can vary widely from packet to packet. As Figure 4.7 illustrates, this is due primarily to the large distance variations of customers from the central office. In one case, if the closest and farthest customers attached to a common optical power splitter are 20 km apart and if

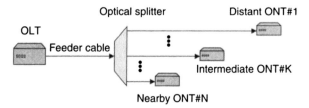

Figure 4.7. Large distance variations of customers from the central office result in different signal power losses across the PON.

the fiber attenuation is 0.5 dB/km, there is a 10-dB difference in the signal amplitudes that arrive at the OLT from these two users if both have the same upstream laser output level. If there is an additional power splitter in the transmission path going to one of the ONTs, the difference in signal levels arriving at the OLT could vary up to 20 dB. Note that such amplitude variations occur only at the OLT and not at an ONT, since a downstream path from an OLT to an ONT is a point-to-point link.

Figure 4.8 shows the consequence of this effect. The top part shows the type of data pattern that would be received in conventional point-to-point links, such as at the customer premises. Here there is no amplitude variation in the logic 1's that are received. The bottom part illustrates the optical signal pattern levels that may arrive

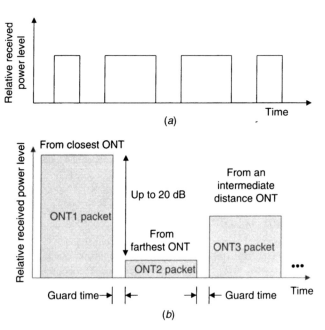

Figure 4.8. (a) Typical data pattern received in conventional point-to-point links; (b) optical signal pattern levels that may arrive at an OLT.

at the OLT. In this case the signal amplitude changes from packet to packet, depending on how far away the ONT is from the central office.

Since a conventional optical receiver is not capable of instantaneous handling of rapidly changing differences in signal amplitude and clock phase alignment, a specially designed *burst-mode receiver* is needed. These receivers can quickly extract the decision threshold and determine the signal phase from a set of overhead bits placed at the beginning of each packet burst. However, this methodology results in a receiver sensitivity power penalty of up to 3 dB.

The key requirements of a burst-mode receiver are high sensitivity, wide dynamic range, and fast response time. The sensitivity is important in relation to the optical power budget (see Chapter 11), since, for example, a 3-dB sensitivity improvement can double the size of the power splitter so that more subscribers can be attached to the PON. A wide dynamic range is essential for achieving a long network reach, that is, to be able to accommodate users located both close and far away from the power splitter.

The use of a conventional ac-coupling method is not possible in a burst-mode receiver, since the residual charge in a coupling capacitor following any particular data burst cannot dissipate fast enough not to affect the initial conditions of the next burst. The burst-mode receiver therefore requires additional circuitry to accommodate dc-coupled operation. Such receivers now are incorporated into standard commercially available OLT equipment.

4.5 BURST-MODE ONT TRANSMISSION

In a conventional optical fiber telecommunication system, a laser in a transmitter normally is biased slightly on to allow the light output to turn on quickly without light signal distortion in response to a electric driving signal. Thus, in general, the ratio of the optical output at the bias point to that of a logic 1 level is around 0.1 (i.e., the bias output level is 10 dB below the signal level).

This biasing procedure is not possible for lasers used by an ONT in a PON. For this case, in the upstream direction no significant level of optical power may be emitted by an ONT laser except during the particular time slot in which the ONT is allowed to transmit. This means that the laser should be turned off almost completely when it is not transmitting data. If this were not done, then as a result of the power-combining nature of the optical splitter in the upstream direction, some residual power from each ONT would be received by the OLT in each time slot. The effect would be a buildup of the noise floor in the OLT receiver to an unacceptable level.

Consequently, in a PON the idle-state output power should be at least 20 to 30 dB below the power emitted during a logic 1 pulse. Since a laser takes a short time to stabilize if it is turned on suddenly from the off state, the various PON specifications allow a time slot of a few bits in advance of a data burst during which a laser transitions to a bias state and stabilizes its output. In addition, a 1-bit time is allowed at the end of a transmitted burst to allow the laser time to turn off. This is done so that after the laser is turned off its output due to the residual tail current does not interfere with the next packet burst from another ONT. Therefore, ONT transmitter circuitry needs to be designed to accommodate *burst-mode laser drivers*.

4.6 PON TRANSCEIVER PACKAGES

Since a PON is based on the use of bidirectional optical fiber links, manufacturers have developed special transceiver configurations that include the light source, the optical receiver functions, and the wavelength multiplexer in a single package. Popular PON transceiver configurations are the *small form factor* (SFF), the *SFF pluggable* (SFP), and the *triple-play* packages. The small package sizes of such transceivers save circuit board space and enable high port densities.

Different options for device speed, laser type (Fabry–Perot or DFB), optical power output, and optical connection are available. There is a single optical interface to the device, which can be either a fiber flylead or an optical connector that is integrated into the package. The term *pigtail* or *flylead* refers to a short length of optical fiber that is attached to a device in the factory. An optical connector can be attached easily to such a flylead for subsequent interfacing to an optical cable. In addition, the OLT needs to have connections for electric signal interfaces to services such as telephony, Internet access, or video. Similarly, the ONT needs to have electric connections for telephone, Internet, and video services to the customer premises.

Figure 4.9 shows a standard SFP transceiver package. Note that it has an optical fiber connector receptacle integrated into the package. This allows easy attachment of an optical connector for coupling to a transmission fiber. A key feature of SFP modules is that they are *hot-pluggable*, which means that one can insert and remove them from the circuit card without turning off the power to the device. This ensures that the equipment never stops running (known as *zero downtime interruption*) when performing online maintenance. The fact that the SFP modules can be exchanged easily also makes it simple to carry out system upgrades to higher speeds or increased capacity, enables rapid link reconfigurations, and helps reduce inventory costs since

Figure 4.9. Standard SFP transceiver package. (Photo courtesy of LuminentOIC; www.LuminentOIC.com.)

Figure 4.10. Triple-play ONT transceiver package that handles three separate services simultaneously. (Photo courtesy of LuminentOIC; www.LuminentOIC.com.)

equipment circuit cards can be stocked without having specific optical transceivers on them.

Figure 4.10 shows a triple-play ONT transceiver package. As described in Chapter 6 the name *triple play* means that it can handle three separate services simultaneously. Downstream voice and data arrive at 1490 nm, upstream voice and data are sent at 1310 nm, and video comes in on a 1550-nm wavelength. Typically, the three wavelengths are sent at different speeds, which is a condition that the transceiver must be able to handle. Later chapters on specific PON types give further information on the various operational categories that have been designated for these types of transceivers.

4.7 SUMMARY

A basic passive optical network uses three separate wavelengths on a single fiber. From the central office an optical source operating at 1490 nm sends combined voice and data traffic to users and a 1550-nm source transmits video content to subscribers. The return link from the users to the central office employs a 1310-nm wavelength for combined voice and data traffic. Typically in the central office the two optical sources are on separate circuit cards which reside in different equipment racks. To reduce circuit board size and since there is only one bidirectional fiber, the optical transmitters and receivers in the ONT equipment at the customer premises often are combined into a single transceiver package. Included in such a package would be optical source modules, photodetection modules, the passive device for combining and separating the individual wavelengths, digital clock and data recovery interfaces for data and voice, alarm detection and laser shutdown circuitry, and video receiver circuitry.

The main laser types applicable to PONs are the Fabry–Perot (FP) laser and the distributed-feedback (DFB) laser. Key properties of these lasers include high optical output powers, narrow linewidths, and highly directional output beams for efficient coupling of light into fibers. For PON applications, these light sources operate in a direct modulation mode, which is the process of using a varying electrical signal to change the optical output level.

For PON applications, the operational characteristics of an OLT optical receiver differ significantly from those used in conventional point-to-point links. This arises from the fact that the amplitude and phase of information packets received in successive time slots from different user locations can vary widely from packet to packet. This is due primarily to the large distance variations of customers from the central office. Since a conventional optical receiver is not capable of handling such rapidly changing differences in signal amplitude, a specially designed burst-mode receiver is needed. Such receivers now are incorporated into standard commercially available OLT equipment.

Since a PON is based on the use of bidirectional optical fiber links, manufacturers have developed special transceiver configurations that include the light source, the optical receiver functions, and the wavelength multiplexer in a single package. Popular PON transceiver configurations are the small form factor (SFF), the SFF pluggable (SFP), and the triple-play packages. The small package sizes of such transceivers save circuit board space and enable high port densities.

PROBLEMS

4.1 The Fabry–Perot (FP) laser shown in Figure 4.1 emits light in a number of longitudinal modes over a broad spectrum. The number of modes depends on the cavity length and their separation is given by $\Delta\lambda = \lambda^2/2nL$.

(a) What is the wavelength separation of the modes for a 1550-nm InGaAsP FP laser if $L = 400\,\mu m$ and the refractive index of the laser material is $n = 3.63$?

(b) How many oscillating modes does this FP laser emit if the device gain is such that the modes are contained in the spectral range 1540 to 1560 nm?

(c) What is the corresponding mode separation in terms of frequency?

4.2 The DFB laser shown in Figure 4.2 goes into a lasing condition at a wavelength λ_B when the Bragg condition $\lambda_B = 2\Lambda n$ is met for the internal grating. Here Λ is the grating period and n is the refractive index of the laser material. What is the grating period for an InGaAsP DFB laser that operates at $\lambda_B = 1550$ nm if the refractive index of the laser material is $n = 3.63$?

4.3 Due to aging factors, the output power of a typical laser diode will decrease exponentially with time according to the relationship $P(t) = P_0 \exp(-t/\tau)$, where P_0 is the initial output power at $t = 0$ and τ is a lifetime factor. Suppose that based on lifetime tests it is predicted that a certain laser type will have its power decreased to 90 percent of its initial value in three years. In how many years will the output be at 50 percent of its initial value?

4.4 The optical power output P_{out} from a laser cavity can be approximated by the relationship $P_{out} = P_0 (1 - R)/[2\alpha L + \log(1/R)]$, where P_0 is a constant, $R < 1$ is the reflectivity of the front laser facet, and $2\alpha L$ is the round-trip loss in the cavity. If the cavity loss is $2\alpha L = 0.7$, plot P_{out}/P_0 as a function of R for values of $0.75 \leq R \leq 1.0$ to show that the output power is a maximum for a reflectivity value of $R = 0.875$.

4.5 Using Eq. (2.9), calculate the loss due to interface reflections when light emitted from an InGaAsP laser (with refractive index 3.63) goes through an airgap ($n = 1.00$ for air) into an optical fiber for which the refractive index is 1.48.

4.6 Using vendor resources found on the Internet, compare the performance parameters of DFB and FP laser diodes operating at the three FTTP wavelengths (1310, 1490, and 1550 nm). List characteristics such as threshold current, fiber-coupled output power, peak emission wavelength, spectral width, rise and fall times, package type, and operating temperature range (consider both cooled and uncooled designs).

4.7 Suppose that 6×10^6 photons at 1310 nm fall on a photodetector in a given optical pulse.

(a) What is the quantum efficiency if on the average 5.4×10^6 electron–hole pairs are generated in the photodetector by these photons?

(b) What is the responsivity of the photodetector?

(c) If the optical power level in this pulse is $10\,\mu W$, what photocurrent is generated?

4.8 An InGaAs avalanche photodiode has a quantum efficiency of 0.90 at 1550 nm. Suppose that 100 nW of optical power produces a multiplied photocurrent of $2.80\,\mu A$. What is the avalanche multiplication factor?

4.9 An LED operating at 1300 nm injects $2.5\,\mu W$ of optical power into an optical fiber. If the fiber attenuation between the LED and the input to a photodetector is 30 dB and the photodetector quantum efficiency is 0.65, what is the probability that fewer than five electron–hole (eh) pairs will be generated in a 1-ns interval? Note that the average number N_{ave} of eh pairs generated in a time interval T is equal to the quantum efficiency times the ratio of the energy received in this time period to the energy per photon. The actual number n of eh pairs generated fluctuates from the average according to the Poisson probability distribution

$$P_r(n) = N_{ave}^n \frac{\exp(-N_{ave})}{n!}$$

FURTHER READING

1. G. Keiser, *Optical Communications Essentials*, McGraw-Hill, New York, 2003.
2. D. Neamen, *Semiconductor Physics and Devices*, 3rd ed., McGraw-Hill, New York, 2003.

3. P. Ossieur, X.-Z. Qiu, J. Bauwelinck, and J. Vandewege, "Sensitivity penalty calculation for burst-mode receivers using avalanche photodiodes," *J. Lightwave Technol.*, vol. 21, pp. 2565–2575, Nov. 2003.
4. E. Säckinger, *Broadband Circuits for Optical Fiber Communication*, Wiley, Hoboken, NJ, 2005.
5. C. Su, L.-K. Chen, and K.-W. Cheung, "Theory of burst-mode receiver and its applications in optical multiaccess networks," *J. Lightwave Technol.*, vol. 15, pp. 590–606, Apr. 1997.
6. TIA/EIA-526-19, "OFSTP-19: Optical Signal-to-Noise Ratio Measurement Procedure for DWDM Systems," Oct. 2000.

5

PASSIVE OPTICAL COMPONENTS

In addition to fibers, light sources, and photodetectors, many other components are used in a complex optical communication network to split, route, process, or otherwise manipulate light signals. The devices can be categorized broadly as either passive or active components. *Passive optical components* play important roles in PONs, since they do not require an external source of energy to perform an operation or transformation on an optical signal. For example, a passive optical filter will allow only a certain wavelength to pass through it while absorbing or reflecting all other light, and an optical splitter divides the light entering it into two or more lower-level optical power streams. *Active components* require some type of external energy either to perform their functions or to be used over a wider operating range than a passive device, thereby offering greater flexibility. However, these devices typically are not used between endpoints in a PON, except within the end equipment. In this chapter we give details on passive devices and fiber cables used for PONs.

Some basic passive functions and the devices that enable them are as follows:

- Transferring light signals: optical fiber
- Attenuating light signals: optical attenuator, isolator
- Influencing the spatial distribution of a lightwave: directional coupler, star coupler, beam expander
- Modifying the state of polarization: polarizer, half-wave plate, Faraday rotator
- Redirecting light: circulator, mirror, grating
- Reflecting light: fiber Bragg grating, mirror
- Selecting a narrow spectrum of light: optical filter, grating
- Converting lightwave modes: fiber gratings, Mach–Zehnder interferometer

---

FTTX Concepts and Applications, by Gerd Keiser
Copyright © 2006 John Wiley & Sons, Inc.

- Combining or separating independent signals at different wavelengths: WDM device
- Providing lightpath continuity from one fiber to another: optical connector and splice

The passive components described in this chapter for PON applications include wavelength-selective couplers, optical power splitters, optical connectors, optical splices, and cables.

5.1 WDM COUPLERS FOR PONS

A fundamental WDM component is the *wavelength multiplexer*. As Figure 5.1 shows, in one direction of operation (left to right in the figure) the function of this device is to combine independent signal streams operating at different wavelengths onto the same fiber. Note that the signal rates and formats can be different on each wavelength. In the reverse direction (right to left in the figure) this device separates the aggregate of incoming wavelengths into individual wavelength channels. Again, data rates and formats can be selected independently on each wavelength. Many techniques using specialized components have been devised for combining multiple wavelengths onto the same fiber and separating them at the receiver. Each of these techniques has certain advantages and various limitations. These techniques include thin-film filters, planar lightwave circuits based on Mach–Zehnder interferometry, arrayed waveguide gratings, fiber Bragg gratings, and diffraction gratings.

Prior to 2000 the standard wavelength spacing was 100 GHz for 2.5-Gbps DWDM links. Subsequently, high-speed telecommunication applications using many wavelengths moved toward 10-Gbps ultradense systems operating with channels spaced 50, 25, or 12.5 GHz apart. For example, at 10-Gbps transmission rates in the C-band, it is possible to send 96 channels that are separated by 50 GHz. The expansion of DWDM channels beyond the C-band into the S- and L-bands has allowed the possibility of sending 320 wavelengths spaced 25 GHz apart in the combined C- and L-band with 10-Gbps transmission rates per channel.

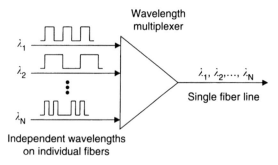

Figure 5.1. Basic concept of wavelength-division multiplexing.

In contrast to the tightly packed DWDM channels used for high-capacity telecommunication links, a PON is based on CWDM. Three specific wavelengths nominally are used in a PON, such as a fiber-to-the-premises (FTTP) network, to carry optical signals associated with different types of services. These wavelengths are 1310, 1490, and 1550 nm. To combine or separate the wavelengths into different transmit or receive channels, one uses a wavelength-selective coupler. Such a combining or separating device could be either a 1 × 2 or a 1 × 3 coupler (see Chapter 3).

Wavelength multiplexers for CWDM applications have less stringent performance demands than DWDM has for certain parameters, such as center wavelength stability tolerance, its change with temperature, and the spectral passband sharpness. However, they still need to have a good reflection isolation, a small polarization-dependent loss, and low insertion losses. Two technologies for making CWDM devices for PON applications are thin-film filters and transmission diffraction gratings. Note that special packaging designs can make these devices be athermal, so that no active temperature control is needed to maintain their stability with temperature changes. Other WDM multiplexer technologies, such as arrayed waveguide gratings, fiber Bragg gratings, and reflection diffraction gratings are more applicable to DWDM applications.

5.1.1 Thin-Film Filters

A dielectric *thin-film filter* (TFF) is an *optical bandpass filter*. This means that it allows a particular design-specific wavelength band to pass straight through it and reflects all other light. Depending on the TFF device design, this passband can be made as narrow as 50 GHz (or equivalently, a 0.5-dB spectral band of 0.2 nm at a 1550-nm wavelength). Alternatively, a TFF can be used as a band splitter to separate two wavelength regions. Figure 5.2 shows an example of a 1 × 2 band-splitting component that can be used in bidirectional PON applications. Such a compact device

Figure 5.2. A 1 × 2 band splitting device for bidirectional PON applications. (Photo courtesy of Lightwaves2020; www.lightwaves2020.com.)

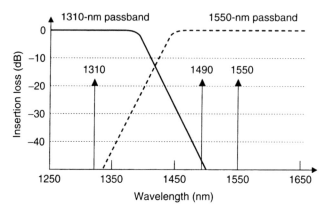

Figure 5.3. A passband coupler can provide better than 45-dB isolation between the 1310-nm upstream channel and the downstream wavelengths.

has a typical length of 40 mm and a 5.5-mm diameter. As Figure 5.3 shows, for a PON application this type of coupler provides a high isolation (better than 45 dB) between the 1310-nm upstream channel and the 1490- and 1550-nm downstream wavelengths. Figure 5.4 illustrates its application at a customer premises, for example. Here a 1310-nm upstream signal enters the transmission port on the left and leaves from the common port on the top right that heads toward the central office. Optical signals at 1490 and 1550 nm coming into the common port from the central office are reflected in the device and come out of the reflection port on the bottom right for use by equipment collocated with the coupler.

The basis of TFF devices is analogous to that of the classical Fabry–Perot cavity structure used for laser diodes (see Section 4.1). For a TFF a resonance cavity is formed by means of two parallel reflective surfaces on opposite faces of a thin dielectric film, as shown in Figure 5.5. This structure is called either a *Fabry–Perot interferometer* or an *etalon*. To see how it works, consider a light signal that is incident on the left surface of the etalon. After the light passes through the cavity and hits the inside surface on the right, some of the light leaves the cavity and some is reflected. The amount of light that is reflected depends on the reflectivity R of the surface. If the

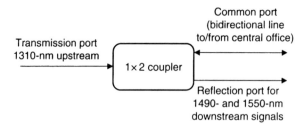

Figure 5.4. Application of a 1 × 2 band-splitting coupler at a customer's premises.

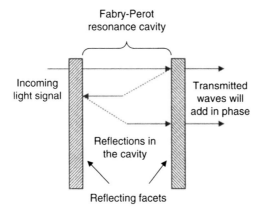

Figure 5.5. Structure of a Fabry–Perot interferometer or an etalon.

round-trip distance between the two mirrors is an integral multiple of a wavelength λ (i.e., λ, 2λ, 3λ, etc.), all light at those wavelengths that pass through the right facet *add in phase*. This means that these wavelengths *interfere constructively* in the device output beam so their intensities add. These wavelengths are called the *resonant wavelengths* of the cavity. The etalon reflects all other wavelengths.

To understand some of the terminology used for a TFF, let us look at some basic etalon theory. The transmission T of an ideal etalon in which there is no light absorption by the mirrors is an Airy function given by

$$T = \left[1 + \frac{4R}{(1-R)^2} \sin^2\left(\frac{\varphi}{2}\right)\right]^{-1} \tag{5.1}$$

where R is the *reflectivity* of the mirrors (the fraction of light reflected by the mirror) and φ is the round-trip phase change of the light beam. If one ignores any phase change at the mirror surface, the phase change for a wavelength λ is

$$\varphi = \frac{2\pi}{\lambda} 2nD \cos\theta \tag{5.2}$$

where n is the refractive index of the dielectric layer that forms the mirror, D the distance between the mirrors, and θ the angle to the normal of the incoming light beam.

Figure 5.6 gives a generalized plot of Eq. (5.1) over the range $-3\pi \leq \varphi \leq 3\pi$. Since φ is proportional to the optical frequency $f = 2\pi/\lambda$, Figure 5.6 shows that the power transfer function T is periodic in f. The peaks, called the *passbands*, occur at those wavelengths that satisfy the condition $N\lambda = 2nD$, where N is an integer. Thus, for a single wavelength to be selected by the filter from a particular spectral range, all the wavelengths must lie between two successive passbands of the filter transfer function. If some wavelengths lie outside this range, the filter would transmit several

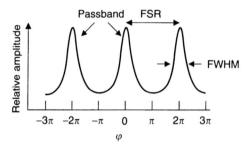

Figure 5.6. Airy function. The distance between adjacent peaks is called the *spectral free range* (FSR).

wavelengths. The distance between adjacent peaks is called the *free spectral range* (FSR). This is given by

$$\text{FSR} = \frac{\lambda^2}{2nD} \tag{5.3}$$

Another important parameter is the measure of the full width of the passband at its half-maximum value, which is designated by FWHM (*full-width half-maximum*). This is of interest in WDM systems for determining how many wavelengths can lie within the FSR of the filter. The ratio FSR/FWHM gives an approximation of the number of wavelengths that a filter can accommodate. This ratio is known as the *finesse F* of the filter and is given by

$$F = \frac{\pi\sqrt{R}}{1-R} \tag{5.4}$$

A typical TFF consists of multilayer thin-film coatings of alternating low- and high-index materials, such as SiO_2 and Ta_2O_5, as shown in Figure 5.7. The layers usually

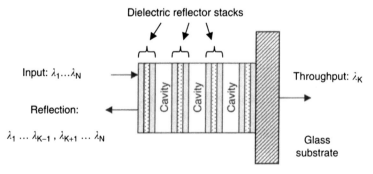

Figure 5.7. A multilayer optical thin-film filter consists of a stack of several dielectric thin films separated by cavities.

are deposited on a glass substrate. Each dielectric layer acts as a nonabsorbing reflecting surface, so that the structure is a series of resonance cavities each of which is surrounded by mirrors. As the number of cavities increases, the passband of the filter sharpens up to create a flat top for the filter, which is a desirable characteristic for a practical filter. In Figure 5.7 the filter is made such that if the input spectrum contains wavelengths λ_1 through λ_N, only λ_k passes through the device. All the other wavelengths are reflected. Thin-film filters are available in a wide range of passbands, varying from 50 to 800 GHz and higher for widely spaced channels.

To create a wavelength multiplexing device for combining or separating N wavelength channels, one needs to cascade $N-1$ thin-film filters. Figure 5.8 illustrates a multiplexing function for the four wavelengths λ_1, λ_2, λ_3, and λ_4. Here the filters labeled TFF_2, TFF_3, and TFF_4 pass wavelengths λ_2, λ_3, and λ_4, respectively, and reflect all others. The filters are set at a slight angle to direct light from one TFF to another. First filter TFF_2 reflects wavelength λ_1 and allows wavelength λ_2 to pass through. These two signals then are reflected from filter TFF_3 where they are joined by wavelength λ_3. After a similar process at filter TFF_4 the four wavelengths can be coupled into a fiber by means of a lens mechanism.

To separate the four wavelengths from one fiber into four physically independent channels, the directions of the arrows in Figure 5.8 are reversed. Since a light beam loses some of its power at each TFF because the filters are not perfect, this multiplexing architecture works for only a limited number of channels. This usually is specified as being 16 channels or less.

Table 5.1 lists typical performance parameters for commercially available wavelength multiplexers based on thin-film filter technology. The parameters address eight-channel DWDM devices with 50- and 100-GHz channel spacings and an eight-channel CWDM module.

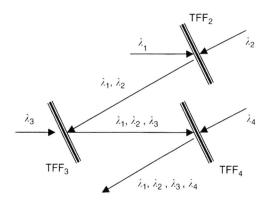

Figure 5.8. Multiplexing four wavelengths using thin-film filters.

TABLE 5.1 Typical Performance Parameters for Eight-Channel DWDM and CWDM Multiplexers Based on Thin-Film-Filter Technology

Parameter	50-GHz DWDM	100-GHz DWDM	20-nm CWDM
Number of channels	8	8	8
Center wavelength accuracy	±0.1 nm	±0.1 nm	±0.3 nm
Channel passband at 0.5-dB bandwidth	±0.20 nm	±0.11 nm	±6.5 nm
Insertion loss	≤4.0 dB	≤4.0 dB	≤4.5 dB
Ripple in passband	≤0.5 dB	≤0.5 dB	≤0.5 dB
Adjacent channel isolation	≥23 dB	≥20 dB	≥15 dB
Directivity	≥50 dB	≥55 dB	≥50 dB
Optical return loss	≥40 dB	≥50 dB	≥45 dB
Polarization dependent loss	≤0.1 dB	≤0.1 dB	≤0.1 dB
Thermal wavelength drift	<0.001 nm/°C	<0.001 nm/°C	<0.003 nm/°C
Optical power capability	500 mW	500 mW	500 mW

5.1.2 Transmission Diffraction Gratings

A *diffraction grating* is a conventional optical device that spatially separates the different wavelengths contained in a beam of light. The device consists of a set of diffracting elements, such as narrow parallel slits or grooves, separated by a distance comparable to the wavelength of light. These diffracting elements can be either reflective or transmitting, thereby forming a reflection grating or a transmission grating, respectively. Separating and combining wavelengths in diffraction gratings is a parallel process, in contrast to the serial process that is used with thin-film filters. A diffraction grating has very low adjacent-channel crosstalk, which usually is less than 30 dB. Insertion loss is typically less than 3 dB and is uniform to within 1 dB over a large number of channels. A passband of 30 GHz at 1-dB ripple is standard.

Figure 5.9 shows the operational concept of a *transmission grating*, which also is known as a *phase grating*. The main characteristic is that it consists of periodic

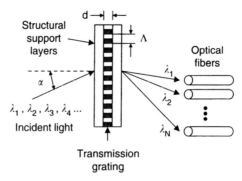

Figure 5.9. Each wavelength emerges at a slightly different angle after passing through a transmission grating.

variations of the refractive index of the grating that have a spacing designated by the parameter Λ. When a spectrum of plane waves is incident on the grating at an angle α_i, the individual wavelengths λ_k in the spectrum will emerge from the grating at slightly different angles $\alpha_e(\lambda_k)$. These angles are determined by the *grating equation*

$$\sin \alpha_e(\lambda_k) = \sin \alpha_i + q \frac{\lambda}{\Lambda} \quad (5.5)$$

where q is the *diffraction order* of the grating. Usually, only the first order ($q = -1$) is considered. If the angle α_i is small, that is, when the wavelengths $\lambda \ll \Lambda$, Eq. (5.5) can be approximated by

$$\alpha_e(\lambda_k) = \alpha_i + q \frac{\lambda}{\Lambda} \quad (5.6)$$

After a spectrum of wavelength channels passes through the grating (from left to right in Figure 5.9), they can be focused individually into distinct receiving fibers since each wavelength emerges at a slightly different angle. The operation of this device is reciprocal. That is, in the reverse direction (from right to left in the figure), the grating focuses the individual wavelengths coming in on the fiber array into the single fiber on the left.

5.2 OPTICAL POWER SPLITTER

An $N \times N$ *star coupler* or *splitter* is a more general form of the 2×2 coupler, as shown in Figure 5.10. This device is a key component in a passive optical network for distributing optical signals to customers from the same feeder fiber (e.g., see Chapter 6). In the broadest application, star couplers combine the light streams from two or more input fibers and divide them among several output fibers. In the general case, the splitting is done uniformly for all wavelengths, so that each of the N outputs receives $1/N$ of the power entering the device.

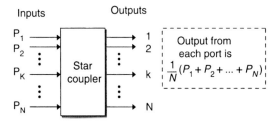

Figure 5.10. An $N \times N$ star coupler or splitter is a more general form of the 2×2 coupler.

5.2.1 Splitting Loss

In an ideal star coupler the optical power from any input is evenly divided among the output ports. The total loss of the device consists of its splitting loss plus the excess loss in each path through the star. The *splitting loss* is given in decibels by

$$\text{splitting loss} = -10 \log \frac{1}{N} = 10 \log N \tag{5.7}$$

Extending the expression for a 2×2 coupler given in Eq. (3.3) to a star coupler, for a single input power P_{in} and N output powers, the excess loss in decibels is given by

$$\text{star coupler excess loss} = 10 \log \frac{P_{in}}{\sum_{i=1}^{N} P_{out,i}} \tag{5.8}$$

The insertion loss and crosstalk can be found from Eqs. (3.4) and (3.5), respectively.

5.2.2 Optical Splitter Structure

High-quality optical splitters can be made by using either *planar lightwave circuit* (PLC) technology or a series of 2×2 fused-fiber single-mode couplers. Figures 5.11 and 5.12 illustrate the structural concepts of using fused-fiber and PLC splitters, respectively, for making a 1×8 optical power splitter. When constructing a splitter using 2×2 fused-fiber couplers, only one input arm is used, as illustrated at the top of Figure 5.11. The PLC-based splitter shown in Figure 5.12 consists of a series of

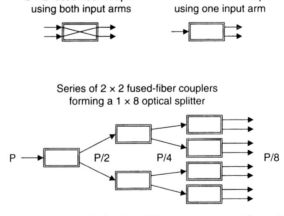

Figure 5.11. Structural concepts of using fused-fiber couplers for making a 1×8 optical power splitter.

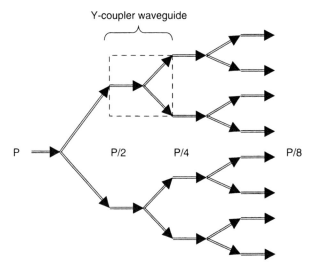

Figure 5.12. Structural concepts of using PLC splitters for making a 1 × 8 optical power splitter.

joined Y-coupler waveguides. The operation of these devices is reciprocal; that is, they can function as either optical power splitters to divide an incoming light signal into $N \leq 32$ branches (nominally for a PON) or as power-combining devices to multiplex optical signal streams from individual fibers onto a common fiber. The link design details of how this works in an FTTP network, for example, are given in Chapter 10. The PLC-based splitters offer more compact packages than fused-fiber splitters for situations where size is an issue. Both types of couplers are available in 1 × 4, 1 × 8, 1 × 16, or 1 × 32 configurations.

Figure 5.13 gives an example of a 1 × 32 PLC splitter. Table 5.2 presents some typical values of the performance parameters for commercially available 1 × 8 and 1 × 32 PLC optical splitters.

Figure 5.13. Packaged 1 × 32 optical power splitter using PLC technology. (Photo courtesy of SENKO Advanced Components; www.senko.com.)

TABLE 5.2 Representative Specifications for 1 × 8 and 1 × 32 PLC Optical Splitters

Parameter	Unit	1 × 8 Specification	1 × 32 Specification
Operating wavelength	nm	1280–1310 or 1480–1600	1280–1360 or 1480–1580
Average insertion loss	dB	10.7	18
Return loss	dB	≥55	≥55
Operating temperature	°C	−40 to +85	−40 to +85
Uniformity	dB	≤1.0	≤2.5
Size (width × height × length)	mm	5 × 4 × 45	7 × 4 × 55

5.3 OPTICAL CABLES FOR PONS

The optical fiber used predominantly for PONs is the G.652c or G.652d single-mode type. Enclosing these optical fibers within some type of cabling structure provides physical and environmental protection. Different cable designs are used depending on where in the PON they are deployed. The cable structure will vary greatly depending on whether the cable is to pulled or blown into underground or intrabuilding ducts, buried directly in the ground, installed on outdoor poles, or placed under water. To prevent excessive stretching, the cabling process usually includes the incorporation of *strength members* into the cable design. This is especially important in the design of aerial cables that can experience severe stresses due to factors such as wind forces or ice loading. Common strength members are strong nylon yarns (e.g., Kevlar), steel wires, and fiberglass rods.

5.3.1 Cable Structures

Two basic structures are the tight-buffered fiber cable design and the loose-tube cable configuration. As shown in Fig. 5.14 for a *simplex* one-fiber cable, in the *tight-buffered cable* design each fiber is individually encapsulated within its own 900-μm-diameter plastic buffer structure. This feature gives the tight-buffered cables excellent moisture and temperature performance and also allows easy attachment of optical

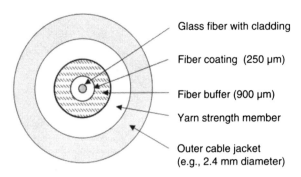

Figure 5.14. Cross section of a simplex tight-buffered fiber cable.

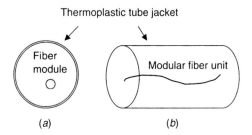

Figure 5.15. Concept of a loose-tube cable construction: (*a*) end view; (*b*) side view.

connectors. Surrounding this 900-μm structure is a layer of cable-strengthening material such as a tough aramid yarn. A polyvinyl chloride (PVC) outer jacket then encapsulates the entire structure.

In the *loose-tube cable* configuration, one or more modular units consisting of standard fibers with a 250-μm buffer coating plus cabling materials such as wrapping tapes and strength members, are enclosed in a thermoplastic tube jacket. The inner diameter of the tube is much larger than the fiber diameter, as shown in Figure 5.15, and generally contains a water-blocking material. The fiber modules in the tube are slightly longer than the cable itself. These cabling conditions isolate the fibers from the cable and enable them to move freely in the tube. This allows the cable to stretch under tensile loads without applying stress on the fibers.

Cables with tight-buffered fibers nominally are used indoors, whereas the loose-tube structure is intended for outdoor applications. Figure 5.16 gives an example of a loose-tube cable that also has a steel armoring layer just inside the jacket to offer crush resistance and protection against gnawing rodents. Such a cable can be used for direct-burial or aerial-based outside plant applications.

To facilitate the field operation of splicing cables containing a large number of fibers, cable designers devised the fiber-ribbon structure. As shown in Fig. 5.17, the *ribbon cable* is an arrangement of fibers that are aligned precisely next to each other and then encapsulated in a plastic buffer or jacket to form a long, continuous ribbon. The number of fibers in an individual ribbon typically ranges from four to 12. These ribbons

Figure 5.16. Loose-tube cable that has a protective steel armoring layer inside the jacket. (Photo courtesy of OFS; www.ofs.com.).

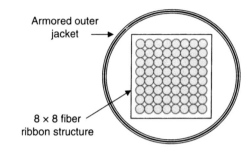

Figure 5.17. Layered 64-fiber ribbon cable.

Figure 5.18. Ribbon cables: (*a*) small cable designed for air-blown installations into ducts; (*b*) lightly armored cable to support aerial, direct-burial, or duct installations. (Photos courtesy of OFS; www.ofs.com.)

can be stacked on top of each other to form a densely packed arrangement of many fibers (e.g., 64 fibers, as illustrated in Figure 5.17) within a cable structure. The ribbon structure also is advantageous in PON field installations when attaching fiber cables to optical power splitters. Figure 5.18 gives two examples of ribbon cables, one of which is designed for air-blown installations into ducts (see Chapter 12) and can contain up to 96 fibers. The other cable has different metallic or dielectric jacketing options to support aerial, direct-burial, or duct installations for fiber counts ranging from 12 to 432.

5.3.2 Fiber and Jacket Color Coding

To distinguish between different fibers in a cable, each fiber is designated by a separate and distinct buffer coating color. The TIA/EIA-598-B standard, *Optical Fiber Cable Color Coding*, prescribes a common set of fiber colors. Since nominally there are up to 12 fiber strands in a single grouping, strands 1 through 12 are uniquely color-coded, as listed in Table 5.3. If there are more than 12 fibers within an individual grouping, strands 13 through 24 repeat the same fundamental color code as those for strands 1 through 12, with the addition of a black or yellow dashed or solid tracer line, as noted in Table 5.3. For cables having more than one tube that contains

TABLE 5.3 Standard Optical Fiber Buffer and Loose-Tube Color Identifications

Fiber Number	Color	Fiber Number	Color
1	Blue	13	Blue/black tracer
2	Orange	14	Orange/yellow tracer
3	Green	15	Green/black tracer
4	Brown	16	Brown/black tracer
5	Slate (gray)	17	Slate/black tracer
6	White	18	White/black tracer
7	Red	19	Red/black tracer
8	Black	20	Black/yellow tracer
9	Yellow	21	Yellow/black tracer
10	Violet	22	Violet/black tracer
11	Rose (pink)	23	Rose/black tracer
12	Aqua	24	Aqua/black tracer

a fiber grouping, the tubes also are color-coded in the same manner as the fibers; that is, tube 1 is blue, tube 2 is orange, and so on.

Ribbon cables follow the same coloring-coding scheme. Thus, one of the outside fibers would have a blue jacket, the next fiber would be orange, and so on, until the other outer edge is reached; for a 12-fiber ribbon this outermost fiber would be aqua (light blue).

5.4 FIBER INTERCONNECTIONS

A significant factor in any fiber optic system installation is the requirement to interconnect fibers in a low-loss manner. These interconnections occur at the optical source, at the photodetector, at intermediate points within a cable where two fibers join, and at intermediate points in a link where two cables are connected. The particular technique selected for joining the fibers depends on whether a permanent bond or an easily demountable connection is desired. A permanent bond (usually within a cable) is referred to as a *splice*, whereas a demountable joint at the end of a cable is known as a *connector*.

5.4.1 Optical Connectors

Many types of optical fiber connectors are available for a variety of applications. Their uses range from simple single-channel fiber-to-fiber connectors in a benign location to rugged multichannel connectors used under the ocean or for harsh military field environments. Connectors are available in designs that screw on, twist on, or snap in place. The snap-on designs are the ones used most commonly for PON applications. The designs include both single- and multichannel assemblies for cable-to-cable and cable-to-circuit card connections.

The majority of connectors use a butt joint coupling mechanism, as illustrated in Figure 5.19. The elements shown in this figure are common to most connectors. The

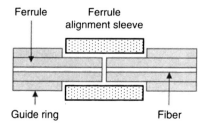

Figure 5.19. Ferrules inserted into a precision butt-joint alignment sleeve.

key components are a long, thin stainless steel, glass, ceramic, or plastic cylinder known as a *ferrule* and a precision sleeve into which the ferrule fits. This sleeve is known variably as an *alignment sleeve*, an *adapter*, or a *coupling receptacle*. The center of the ferrule has a hole that precisely matches the size of the fiber cladding diameter. Typically, the hole size is $125.0 \pm 1.0\,\mu m$. The fiber is secured in the hole with epoxy and the end of the ferrule is polished flat to a smooth finish. A guide ring controls how far the ferrule is inserted into the alignment sleeve. The mechanical challenges in fabricating a ferrule include maintaining both the dimensions of the diameter and the position of the hole relative to the ferrule outer surface.

A connector in which the fiber face is polished perpendicular to the axis of the fiber is called a *straight-polished connector* (SPC). To reduce back reflections at a connector, the fiber end face can be polished at a slight angle. This typically is 8° with respect to the fiber axis. Such a connector is called an *angle-polished connector* (APC).

Most of the ferrules used in optical connectors are made of ceramic material, due to some of the desirable properties they possess. These include the low insertion loss required for optical transmission, remarkable strength, small elasticity coefficient, easy control of product characteristics, and strong resistance to changes in environmental conditions such as temperature.

Further emphasis on reducing connector sizes for higher packaging densities resulted in many concepts for *small form factor* (SFF) single-fiber connector types. The biggest difference among the SFF connectors is whether they use ceramic or plastic ferrules. As their name implies, SFF connectors are small and are designed for fast termination in the field.

Two popular connectors for PONs are the SC and the LC designs. The SC connector was developed by NTT (Nippon Telegraph and Telephone) of Japan. The connector is mated by a simple snap-in method and can be disconnected by pushing in on a tab and pulling off the connector. SC connectors are available in either simplex or duplex configurations. Figure 5.20 shows an example of a simplex single-mode SC connector and its associated adapter. The SC connector uses a precision zirconia (zirconium dioxide) ceramic ferrule. For multimode fibers this yields a typical insertion loss of 0.4 dB when using a manual polishing method or 0.2 dB when using an automated fiber polisher. Single-mode connectors typically achieve a 0.3-dB insertion loss and a 40-dB return loss using a simple manual polishing method. The

Figure 5.20. (a) Simplex single-mode SC connector and (b) its associated adapter. (Photo courtesy of SENKO Advanced Components; www.senko.com.)

mating and loss features allow the connector to be used in tight spaces. An example of such spaces is a patch panel where there is a high packing density of connections.

SC simplex connectors have an outer plastic housing that is color-coded in accordance with TIA/EIA-568-B.3 requirements. This document specifies beige for multimode fibers and blue for single-mode fibers. Duplex SC connectors are used widely on two-fiber patch cords and have been designed with a keying mechanism to maintain fiber optic cabling polarity. The word *polarity* means that the ends of a cable are matched properly for transmit and receive functions. Basically, the duplex configuration combines two standard SC connectors in a common duplex plastic housing. This housing is keyed to maintain fiber polarity and provides smooth insertion and removal of the connector pairs.

Lucent developed the LC connector to meet the growing demand for small, high-density fiber optic connectivity on equipment bays, in distribution panels, and on wall plates. This connector is half the size of an SC connector. Two common and proven technologies were combined in the LC connector. These are the industry-standard RJ-45 telephone plug interface and ceramic-ferrule technology. The advantage of the RJ-45 housing is that it provides a reliable and robust latching mechanism for the LC connector. The LC connector has a six-position tuning feature to achieve very low insertion loss performance by optimizing the alignment of the fiber cores. The insertion loss for an LC connector is typically 0.1 dB.

The LC connectors are available in both simplex and duplex configurations, as shown in Figure 5.21. They are available in industry standard beige and blue colors for multimode and single-mode applications, respectively. The connectors will accommodate 900-μm buffered fiber and 1.60-mm (0.063-in.) and 2.40-mm (0.094-in.) jacketed cable. Duplex and simplex LC adapters with either ceramic or metal sleeves are available in industry standard colors.

5.4.2 Connector Losses

Every joining technique is subject to certain conditions that can cause varying degrees of optical power loss at the joint. These losses depend on factors such as the mechanical alignments of the two fibers, differences in the geometrical and waveguide characteristics of the two fiber ends at the joint, and the fiber end-face qualities.

The most common misalignment occurring in practice, which also causes the greatest power loss, is *axial displacement*. This axial offset reduces the overlap area

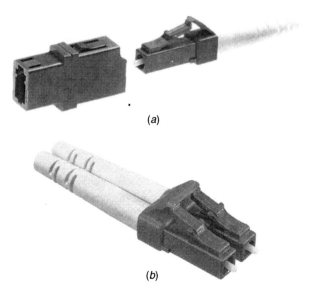

Figure 5.21. LC connectors are available in either (a) simplex or (b) duplex configurations. (Photos courtesy of SENKO Advanced Components; www.senko.com and OFS; www.ofs.com.)

of the two fiber-core end faces, as illustrated in Figure 5.22, and consequently reduces the amount of optical power that can be coupled from one fiber into another. In practice, axial offsets of less than 1 μm are achievable for multimode fibers, which result in losses of less than 0.1 dB. For single-mode fibers, axial offsets of less than 0.1 μm result in losses of less than 0.1 dB.

The derivation of an expression for calculating the axial offset losses in fibers is straightforward. The following equation gives the common core area of two identical step-index fibers of radius a that are axially misaligned by a separation d:

$$A_{common} = 2a^2 \arccos \frac{d}{2a} - da\left[1 - \left(\frac{d}{2a}\right)^2\right]^{1/2} \qquad (5.9)$$

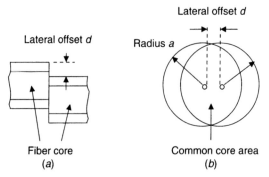

Figure 5.22. Axial offset reduces the common core area of two fiber end faces. (a) lateral view; (b) end view.

The coupling loss for axial offsets then is given by

$$L_{\text{offset}} = -10 \log \frac{A_{\text{common}}}{\pi a^2} \qquad (5.10)$$

5.4.3 Optical Splices

A fiber *splice* is a low-loss bond between two fibers, which can be made using either fusion splicing or mechanical splicing. In making and evaluating optical fiber splices, one must take into account the physical differences in the two fibers, fiber misalignments at the joint, and the mechanical strength of the splice.

The physical differences in fibers that lead to splice losses result in what is called *intrinsic loss*. These fiber-related differences include variations in core diameter, core-area ellipticity, and core–cladding concentricity of each fiber. *Extrinsic losses* depend on how well the fibers are prepared and the care taken to make the splice. The factors here include fiber misalignments at the joint, the smoothness and cleanliness of the fiber end faces, and the skill of the splice-equipment operator. When the fiber bonding is done properly using high-quality equipment, the total splice loss typically is 0.05 to 0.10 dB for fusion splicing and around 0.5 dB for mechanical splices.

Fusion splices are made by thermally bonding together prepared fiber ends. The bonding is done with either an electric arc or a laser pulse so that the fiber ends are melted momentarily and hence bonded together. This technique can produce very low splice losses (typically, averaging less than 0.1 dB).

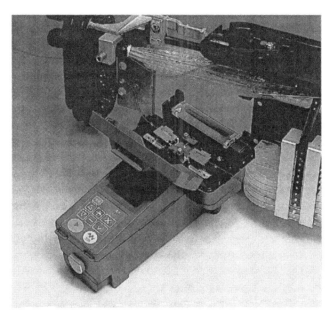

Figure 5.23. Handheld battery-powered fiber splicing instrument for FTTP applications. (Photo of Model S121A courtesy of Furukawa America, Inc.; www.FurukawaAmerica.com.)

Several instrument manufacturers have developed small portable splicing equipment for PON installations. Figure 5.23 shows an example of a compact handheld battery-powered unit. Alternatively, such an instrument can be run with an external 12 V supply using an ac adapter. This unit can splice two fibers in less than 13 seconds with maximum splice losses of less than 0.1 dB. The instrument also has an adjustable 2.5-in. color LCD screen that allows the operator to observe the fiber joint during the splicing process.

The setup in Figure 5.23 shows how the splicing instrument is used to fuse fibers entering a splice holder. After a splice has been made, the slack length of fiber is coiled and stored in a special holding tray shown in the bottom right of the photo. Once all splices are completed between the two cable sections, a protective sealing cover is placed around the splice holder to shield the splices from the environment.

5.5 SUMMARY

Passive optical components obviously play an important role in PONs, since they do not require an external source of energy to perform an operation or transformation on an optical signal. The key passive components for PON applications include wavelength-selective couplers, optical power splitters, optical connectors, and optical splices.

In one direction of operation the function of a wavelength-selective coupler [commonly referred to as a wavelength-division multiplexer (WDM) device] is to combine independent signal streams operating at different wavelengths onto the same fiber. In the reverse direction this device separates the aggregate of wavelengths coming in on one fiber into individual wavelength channels.

In a PON, an optical power splitter is a star coupler that distributes optical signals to customers from the same feeder fiber. In the broadest application, star couplers combine the light streams from two or more input fibers and divide them among N output fibers. In the general case, the splitting is done uniformly for all wavelengths, so that each of the N outputs receives $1/N$ of the power entering the device. High-quality optical splitters can be made by using either planar lightwave circuit technology or a series of 2×2 fused-fiber single-mode couplers.

The optical fiber used predominantly for PONs is the G.652c or G.652d single-mode type. Enclosing these optical fibers within some type of cabling structure provides physical and environmental protection. Different cable designs are used depending on where in the PON they are deployed. To distinguish between different fibers in a cable, each fiber is designated by a separate and distinct buffer color as specified by the TIA/EIA-598 optical fiber cable color-coding standard.

A significant factor in any fiber optic system installation is the requirement to interconnect fibers in a low-loss manner. A permanent bond (usually within a cable) is referred to as a splice, whereas a demountable joint at the end of a cable is known as a connector. Since no signal amplifiers are contained within a PON, these components must exhibit very low loss to meet the stringent power budgets imposed on passive optical networks.

PROBLEMS

5.1 Using product data sheets from vendors such as those available via the Internet, compare the performance characteristics of various offerings of thin-film filters. Consider bandpass ranges from 50 to 400 GHz.

5.2 Search through vendor data sheets to find TFF devices that are appropriate for the setup shown in Figure 5.4.

5.3 Plot Eq. (5.1) as a function of wavelength over the range $1500\,\text{nm} \le \lambda \le 1600\,\text{nm}$. Use the following parameter values: $R = 0.99$, $D = 31\,\mu\text{m}$, $n = 1.50$, and $\theta = 0$. Show that the FSR is 26 nm.

5.4 (a) Consider a TFF that has the parameter values $R = 0.99$, $D = 31\,\mu\text{m}$, and $n = 1.50$. What is the FSR at 1550 nm?
(b) If the spacing between DWDM channels is 0.8 nm, how many wavelength channels fit into the FSR of this TFF?
(c) What is the finesse of the filter?

5.5 If the optical power of each input wavelength shown in Figure 5.8 is 1 mW and the insertion loss of each TFF is 1.0 dB, what are the power levels of the four wavelengths emerging from the final thin-film filter TFF_4?

5.6 A first-order bulk transmission grating is made with a groove frequency of 1100 grooves/mm.
(a) What is the form of the grating equation for such a device?
(b) If the angle of the incident beam is 30°, plot the diffraction angles for the wavelength range $1530\,\text{nm} \le \lambda \le 1560\,\text{nm}$.

5.7 Suppose that an $N \times N$ star coupler is constructed of n 3-dB 2×2 couplers, each of which has a 0.1-dB insertion loss. Find the maximum value of n and the maximum size N if the power budget for the star coupler is 16 dB.

5.8 Suppose that the optical power emerging from a feeder fiber is to be distributed among eight individual houses. Assume that they are separated by 100 m and lie along a straight line going out from the end of the feeder fiber. One way to distribute the power is to use a 1×8 star coupler and run individual fibers to each house. Why is this preferable to the configuration where a single fiber runs along the line to the homes and individual tap couplers in the fiber line each extract 10 percent of the power from the line for each house that the line passes?

5.9 Derive Eq. (5.9), which gives the common core area of two identical step-index fibers of radius a that are axially misaligned by a separation d (see Figure 5.22).

5.10 (a) From Eq. (5.9), what is the loss for identical step-index multimode fibers with radius $a = 25\,\mu\text{m}$ if the axial offset $d = 1\,\mu\text{m}$?
(b) What is the axial offset loss for identical step-index single-mode fibers with radius $a = 4.5\,\mu\text{m}$ if the axial offset $d = 0.3\,\mu\text{m}$?

FURTHER READING

1. A. K. Dutta, N. K. Dutta, and M. Fujiwara, eds., *WDM Technologies: Passive Optical Components*, Elsevier, New York, 2003.
2. R. Freeman, *Fiber Optic Systems for Telecommunications*, Wiley, Hoboken, NJ, 2002.
3. K. Iizuka, *Elements of Photonics*, Vol. II, *For Fiber and Integrated Optics*, Wiley, Hoboken, NJ, 2002.
4. J. Jiang, J. J. Pan, J. Guo, and G. Keiser, "Model for analyzing manufacturing-induced internal stresses in 50-GHz DWDM multilayer thin-film filters and their effects on optical performances," *J. Lightwave Technol.*, vol. 23, pp. 495–503, Feb. 2005.
5. G. Keiser, *Optical Communications Essentials*, McGraw-Hill, New York, 2003.
6. G. Mahlke and P. Gössing, *Fiber Optic Cables: Fundamentals, Cable Design, System Planning*, 4th ed., Wiley, New York, 2001.
7. B. E. A. Saleh and M. C. Teich, *Fundamentals of Photonics*, Wiley, New York, 1991.
8. TIA/EIA-568-B.3, "Optical Fiber Cabling Components Standard," 2000.
9. TIA/EIA-598-B, "Optical Fiber Cable Color Coding," 2001.

6

PASSIVE OPTICAL NETWORKS

In its most general form, an optical network will contain both active and passive optical elements. Active components can be located at the central office, within termination points at the customer's premises, and in the repeaters, switches, and other equipment located in the transmission path between the central office and the customer. Many of these devices are capable of being reconfigured by either a local or a remote control mechanism. They are used for functions such as coupling light from one fiber to another, redirecting the light signal to another transmission path, splitting the signal into two or more branches, amplifying the optical signal power, and processing information contained in the signal. If active devices are used in the transmission path, they require electrical power to perform their functions. In addition, some type of dynamic status monitoring mechanism is needed to verify that the active devices are working properly. If the status information indicates a pending or actual malfunction, a network operator needs to take action and dispatch a maintenance person to the fielded unit to correct the fault. These factors add cost and complexity to the system operation compared to the case in which no active devices are used between network endpoints.

In contrast to conventional networks, a *passive optical network* (PON) has no active components between the central office and the customer's premises. Instead, only completely passive optical components are placed in the network transmission path to guide the traffic signals contained within specific optical wavelengths. Replacing active devices with passive components provides a significant cost savings in maintenance by eliminating the need to power and manage active components in the outside cable plant. In addition, since the passive devices have no electrical power or signal-processing requirements, they have extremely high *mean time between failures* (MTBFs).

In Section 6.1 we present basic PON architectures and highlight the key active modules located in the central office and at or near the customer. Next, in Section 6.2 we describe the characteristics and operations of these endpoint electrooptical modules. We also give a top-level view of how voice, data, and video traffic flows are

---

FTTX Concepts and Applications, by Gerd Keiser
Copyright © 2006 John Wiley & Sons, Inc.

implemented using different wavelengths. The popular term for sending these three service types is *triple play*. In Section 6.3 we discuss some issues related to controlling traffic flows in the downstream and upstream directions. Section 6.4 covers implementation details on the specific passive photonic components used for PONs. In Section 6.5 we introduce three PON alternatives and a point-to-point Ethernet access network scheme and note their differences. Further individual details on these access network schemes are presented in Chapters 7 through 9. For the purpose of optical power-budget calculations, in Section 6.6 we give highlights of the three classes of optical attenuation ranges that the PON specifications define. The standards organizations that deal with PON implementation schemes are introduced in Section 6.7.

6.1 FUNDAMENTAL PON ARCHITECTURE

As the name implies, a *passive optical network* contains no active optical elements at any intermediate points along the network paths. Figure 6.1 illustrates the basic architecture of a typical PON in which a fiber optic network connects switching equipment in a central office with a number of service subscribers. Examples of telecommunication equipment in the central office that might interface to a PON include public switched telephone network (PSTN) switches, Internet protocol (IP) routers, video-on-demand servers, Ethernet switches, asynchronous transfer mode (ATM) switches, and backup storage systems consisting of units such as high-capacity disk arrays and tape-drive libraries.

Starting at the central office, one single-mode optical fiber strand runs to a passive *optical power splitter* near a housing complex, a large apartment or office building, a business park, or some other campus environment. At this point there is a splitting device that simply divides the optical power into N separate paths to the subscribers. If the splitter is designed to divide the incident optical power evenly and if P is the optical power entering the splitter, the power level going to each subscriber

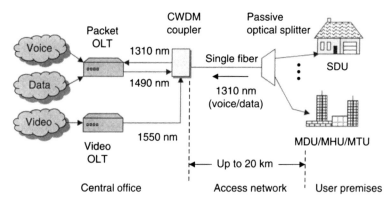

Figure 6.1. Basic architecture of a typical PON.

is *P/N*. Designs of power dividers with other splitting ratios are also possible and there could be more than one splitter in a particular path, depending on the application. The number of splitting paths can vary from 2 to 64, but typically, they are 8, 16, or 32. From the optical splitter, individual single-mode fibers then run to each building or serving equipment. The optical fiber transmission span from the central office to the user can be up to 20 km. In such a network, active devices exist only in the central office and at the end terminal.

The active modules in the network consist of an *optical line terminal* (OLT) situated at the central office and either an *optical network terminal* (ONT) or an *optical network unit* (ONU) at the far end of the network. As shown in Figure 6.2, an ONT is used when the fiber extends into the customer premises, whereas an ONU is used when the fiber line terminates in some type of telecommunication cabinet located near a cluster of homes or businesses. Connections from the ONU to the premises can be by means of media such as twisted-pair wires (e.g., telephone lines or digital subscriber links) or coaxial cable.

For some situations it may be advantageous from a cost-savings point of view to run a single fiber line from the main optical splitter to a distant localized cluster of homes or small businesses or within a centralized location in a neighborhood. In this case, as shown in Figure 6.2, a small optical splitter is located at the end of this fiber line and then short links run from this point to the individual user premises. In contrast to running a long fiber link from the main splitter to each remote customer, this layout lowers the overall network deployment costs significantly. The term *optical distribution network* (ODN) refers to the collection of fibers and passive optical splitters or couplers that lie between the OLT and the various ONTs and ONUs.

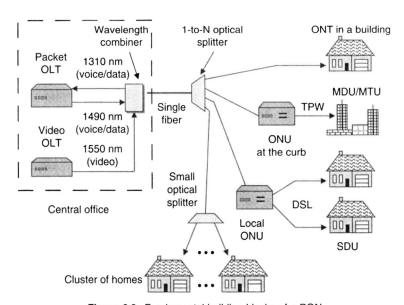

Figure 6.2. Fundamental building blocks of a PON.

The link connecting the central office and the optical splitter is known as a *feeder cable*. One optical splitter may serve up to 32 subscribers. Typically, it is located about 10 km (30,000 ft) from the central office or within 1 km (3000 ft) from the subscribers in a housing neighborhood, a business park, or some other campus. The *distribution cables* originate at the optical splitter. From there they either connect directly to the users or run in a multiple-fiber cable to a local splice box called an *access terminal*. Starting at this splice box, individual *drop cables* connect to the customers' premises.

6.2 ACTIVE PON MODULES

In this section we give a snapshot of the basic functions and compositions of the OLT, ONT, and ONU equipment. Note that for simplicity of discussion, we use the acronym ONT to mean either an ONT or an ONU, unless there is a need to distinguish between the two.

6.2.1 Optical Line Terminal

The OLT is located in a central office and controls the bidirectional flow of information across the ODN. An OLT must be able to support transmission distances across the ODN of up to 20 km. In the downstream direction the function of an OLT is to take in voice, data, and video traffic from a long-haul or metro network and broadcast it to all the ONT modules on the ODN. In the reverse direction (upstream), an OLT accepts and distributes multiple types of voice and data traffic from the network users.

A typical OLT is designed to control more than one PON. Figure 6.3 gives an example of an OLT that is capable of serving four independent passive optical networks. In this case, if there are 32 connections to each PON, the OLT can distribute information to 128 ONTs. As described in Section 6.6, OLT equipment must adhere to specific PON standards, so it can interface with ONT modules from different manufacturers. In addition, the OLT typically is located within a central office. There the environment is fairly stable compared to that of an ONT, which could be housed in

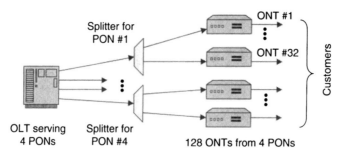

Figure 6.3. OLT that is capable of serving four independent passive optical networks.

an outdoor cabinet or a small enclosure attached to the side of a building. However, an ONT also could be located in a more benign indoor environment.

Simultaneous transmission of separate service types on the same fiber in the ODN is enabled by using different wavelengths for each direction. For downstream transmissions, a PON uses a 1490-nm wavelength for combined voice and data traffic and a 1550-nm wavelength for video distribution. Upstream voice and data traffic use a 1310-nm wavelength. Passive WDM couplers perform the wavelength combining and separation functions, as described in Section 6.4.4. Depending on the particular PON standard being used, the downstream and upstream transmission equipment operates at 155 Mbps, 622 Mbps, 1.25 Gbps, or 2.5 Gbps. In some cases the transmission rates are the same in either direction (a *symmetric* network). In other PON standards the downstream rate may be higher than the upstream rate, which is called an *asymmetric* implementation. A number of different transmission formats can be used for the downstream video transmission at 1550 nm.

6.2.2 Optical Network Terminal

As shown in Figure 6.2, an ONT is located directly at the customer's premises. There its purpose is to provide an optical connection to the PON on the upstream side and to interface electrically to the customer equipment on the other side. Depending on the communication requirements of the customer or block of users, the ONT typically supports a mix of telecommunication services, including various Ethernet rates, T1 or E1 (1.544 or 2.048 Mbps) and DS3 or E3 (44.736 or 34.368 Mbps) telephone connections, ATM interfaces (155 Mbps), and digital and analog video formats.

A wide variety of ONT functional designs and chassis configurations are available to accommodate the needs of various levels of demand. The size of an ONT can range from a simple box that may be attached to the outside of a house to a fairly sophisticated unit mounted in a standard indoor electronics rack for use in large MDU or MTU applications, such as apartment complexes or office buildings. At the high-performance end, an ONT can aggregate, groom, and transport various types of information traffic coming from the user site and send it upstream over a single-fiber PON infrastructure. The term *grooming* means that the switching equipment looks inside a time-division-multiplexed data stream, identifies the destinations of the individual multiplexed channels, and then reorganizes the channels so that they can be delivered efficiently to their destinations.

In conjunction with the OLT, an ONT allows dynamic bandwidth allocation to enable smooth delivery of data traffic that typically arrives in bursts from the users. More details on this allocation function are given in Chapter 7.

6.2.3 Optical Network Unit

An ONU normally is housed in an outdoor equipment shelter. These installations include shelters located at a curb or in a centralized place within an office park. Thus, the ONU equipment must be environmentally rugged to withstand large temperature variations. The shelter for the outdoor ONU must be water-resistant, vandal-proof,

106 PASSIVE OPTICAL NETWORKS

Figure 6.4. Outdoor shelter with a view of the equipment contained inside. (Photo courtesy of ADC; www.adc.com.)

and be able to endure high winds. In addition, there has to be a local power source to run the equipment, together with emergency battery backup. Figure 6.4 shows an outdoor shelter with a view of the equipment contained inside. The link from the ONU to the customer's premises can be a twisted-pair copper wire, a coaxial cable, an independent optical fiber link, or a wireless connection.

6.3 TRAFFIC FLOWS

Two key network functions of an OLT are to control user traffic and to assign bandwidth dynamically to the ONT modules. Since up to 32 ONTs use the same wavelength and share a common optical fiber transmission line, some type of transmission synchronization must be used to avoid collisions between traffic coming from different ONTs. The simplest method is to use *time-division multiple access* (TDMA), wherein each user transmits information within a specific assigned time slot at a prearranged data rate. However, this does not make efficient use of the bandwidth available since many time slots will be empty when several network users do not have information to be sent back to the central office.

  A more efficient process is *dynamic bandwidth allocation* (DBA), wherein time slots of an idle or low-utilization user are assigned to a more active customer. In Chapter 7 we give further details of one specific DBA method. The exact DBA scheme implemented through an OLT in a particular network depends on factors such as user priorities, the quality of service guaranteed to specific customers, the desired response time for bandwidth allocation, and the amount of bandwidth requested (and paid for) by a customer.

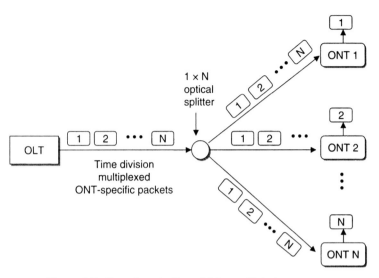

Figure 6.5. Operation of a time-division multiplexing process.

As shown in Figure 6.5, the OLT uses *time-division multiplexing* (TDM) to combine incoming voice and data streams that are destined for users on the PON. As a simple example of this, if there are N independent information streams coming into the OLT, each of which is running at a data rate of R bits per second (bps), the TDM scheme interleaves them electrically into a single information stream operating at a higher rate of $N \times R$ bps. The multiplexed downstream signal is broadcast to all the ONTs. Each ONT discards or accepts the incoming information packets, depending on the packet header addressing. Encryption may be necessary to maintain privacy, since the downstream signal is broadcast and every ONT receives all the information destined for each end terminal.

Sending traffic in the upstream direction is more complicated, since all users have to time share the same wavelength. To avoid collisions between the transmissions of different users, the system uses a TDMA protocol. Figure 6.6 gives a simple example

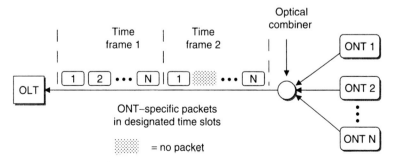

Figure 6.6. Operation of a time-division multiple access protocol.

of this. The OLT controls and coordinates the traffic from each ONT by sending permissions to them to transmit during a specific time slot. The time slots are synchronized so that transmission bursts from different users do not collide. Since each end terminal is located at different distances from the central office, the OLT uses a ranging technique to measure the logical distance between the users and the OLT. This enables each ONT to adjust its transmission timing properly to avoid traffic collisions.

6.4 PASSIVE COMPONENT APPLICATIONS

The three major passive photonic components for PONs are the optical cable types used in the network, the light power splitters within the ODN, and the wavelength couplers use in transceivers.

6.4.1 Optical Cables for PONs

As described in Chapter 12, a wide variety of fiber optic cables may be used in the feeder and distribution sections of a PON. Loose-tube cables for these sections are available with fiber counts ranging from 12 to 432 (see Section 5.3). The fiber packing configuration can consist of either modular groupings of individually jacketed fibers or a stack of fiber ribbons may be used. This type of cable has a long deployment history in telecommunication networks and offers reliability, durability, and ease of installation and routing. As we saw in Figure 5.17, a ribbon cable structure allows the construction of a compact cable for miniature duct installations in the feeder and distribution sections of a PON. Such a cable is much smaller than a conventional loose tube cable and may be installed with the air-jetting process described in Chapter 12. Typical fiber counts range from 8 to 72 and may be as high as 432. In Chapter 10 we give more details on the use of such cables in conjunction with subducts for FTTP applications.

Ribbon and loose-tube cable structures can be used for the optical feeder lines running from the central office to the optical splitter in a PON. Ribbon cables also are used as distribution cables but accessing individual fibers within the ribbon is more labor intensive than is the case for loose-tube cables. Drop cables typically contain one or two fibers for residential interfaces and 6 to 12 fibers for commercial applications in a PON. The design and management of cables for the PON drop section are very important since they need to blend into the neighborhood environment, they must be easy to install, and they need to have high reliability to minimize costs for maintenance testing and troubleshooting procedures.

6.4.2 Optical Power Splitters

The PLC-based optical splitters described in Section 5.2 are popular for PON applications. These are highly reliable components that exhibit stable operation over wide temperature and wavelength ranges. Standard values of these ranges are -40 to $+85°C$ and 1280 to 1650 nm, respectively. Their insertion loss is low

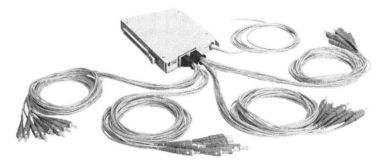

Figure 6.7. Physical configuration of a typical 1 × 32 coupler. (Photo courtesy of ADC; www.adc.com.)

and ranges from 10.5 dB for 1 × 8 couplers to 18 dB for a 1 × 32 device. The loss variation between output ports of such splitters is <1.0 dB for 1 × 8 couplers and <2.2 dB for 1 × 32 couplers. Figure 6.7 illustrates the physical configuration of a typical 1 × 32 coupler. The particular model shown in the figure has short fiber lengths with connectors attached to the splitter outputs for easy attachment to distribution fibers.

To create a compact optical power distribution device, manufacturers use eight-fiber ribbon modules to attach to the internal array of output splitter ports. External to the device, these ribbons are fanned out, and low-loss connectors, such as the SC or LC types, can be attached. As Figure 6.8 illustrates, the connectors may terminate in a miniature patch panel for easy interface to a distribution cable. Alternatively, the fibers in a distribution cable can be spliced to the optical splitter output fibers.

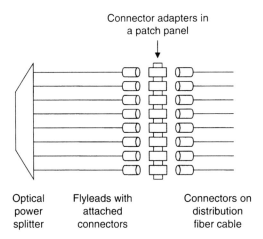

Figure 6.8. Optical power splitter having eight fiber flyleads which interface with a connector patch panel.

However, the use of splices reduces the network flexibility and increases maintenance and troubleshooting testing compared to the use of connectors.

6.4.3 Splitter Enclosures

The optical splitter and its associated connectors or splices need to be housed in an appropriate enclosure to protect them from the particular environment in which they are located. Sealed enclosures are required for all below-surface installations to prevent water ingress. Such enclosures also are suitable for aerial applications. Lower-cost breathable enclosures that prevent water intrusion but are not sealed can be used only in aerial or abovegrade installations. Some illustrations of this hardware are given in Chapter 12.

6.4.4 Wavelength Couplers

Let us look at the specific types and functions of wavelength couplers that can be used in the PON transceivers. A straightforward solution is to use thin-film filters to combine and separate the triple-play wavelengths at the OLT and the ONT. Consider two different TFFs having the following passband characteristics: (1) 1520 to 1600 nm for TFF_1 and 1464 to 1618 nm for TFF_2. Thus, TFF_1 allows a 1550-nm signal to pass through and reflects another at 1490 nm. Similarly, TFF_2 allows the 1550 and 1490 wavelengths to pass through and reflects a signal at 1310 nm. Such devices can be obtained in packages such as the one illustrated in Figure 5.2.

At an OLT the filters can be arranged as shown in Figure 6.9. Here the light output from a 1550-nm laser passes first through TFF_1 and then through TFF_2, since both of these TFFs allow a 1550-nm wavelength to pass through. After exiting TFF_2 the 1550-nm light is coupled into a fiber in the downstream direction. The light from a 1490-nm laser is reflected off TFF_1 and then passes through TFF_2. Similarly, after exiting TFF_2 the 1490-nm light is coupled into a fiber in the downstream direction. After the upstream 1310-nm light enters the OLT transceiver, TFF_2 reflects it into a photodetector.

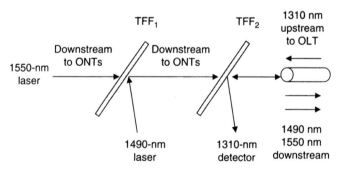

Figure 6.9. Possible arrangement of optical thin-film filters at an OLT for triple-play applications.

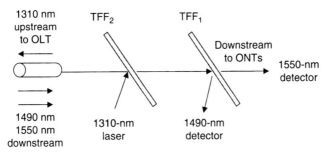

Figure 6.10. Possible arrangement of optical thin-film filters at an ONT for triple-play applications.

The same TFF arrangement can be used at an ONT (or ONU), as shown in Figure 6.10. Here 1310-nm light carrying data and voice traffic aggregated by the ONT is reflected by TFF_2 into an optical fiber in the upstream direction. After the 1490- and 1550-nm information streams come out of the fiber, they first pass through TFF_2. Then the 1490-nm combined voice and data signal is reflected off TFF_1 into a digital receiver and the 1550-nm stream passes through this filter to a video receiver.

6.5 PON ALTERNATIVES

There are several alternative PON implementation schemes, the three main ones being broadband PON (BPON), Ethernet PON (EPON), and gigabit PON (GPON). Table 6.1 lists some of the characteristics of each methodology and the standards to which they adhere. All of these follow the basic PON architecture shown in Figure 6.1. The key differences between them lie in the transmission protocols that are employed. An architecture consisting of point-to-point Ethernet links is a competing access network technology. In this section we introduce these alternatives and note their differences. Further details on BPON, EPON, and GPON are provided in Chapters 7 through 9, respectively.

TABLE 6.1 Major PON Technologies and Some of Their Characteristics

Characteristic	Passive Optical Network Type		
	BPON	EPON	GPON
Standards family	ITU-T G.983	IEEE 802.3ah	ITU-T G.984
Protocol	ATM	Ethernet	ATM and Ethernet
Transmission speeds (Mbps)	622/1244 downstream, 155/622 upstream	1244 upstream, 1244 downstream	155 to 2488 upstream, 1244 or 2488 downstream
Span	20 km	10 km	20 km
Number of splits	32	16 nominal, 32 allowed	64

6.5.1 BPON Basics

Broadband passive optical network standards are based on the G.983 series of ITU-T Recommendations that specify ATM as the transport and signaling protocol (the functions of the ITU-T are described in Section 6.7). In addition, occasionally there will be reference in the literature to an ATM PON (APON). This was the initial PON technology for which standards were established (G.983.1) and now is a subset of the expanded BPON category. ATM is a high-performance switching and multiplexing technology that utilizes fixed-length packets to carry different types of traffic. A key attraction of ATM is that it enables carriers to offer multiple classes of service over metro and wide area networks, to connect devices that operate at different speeds, and to mix a variety of traffic types having different transmission requirements, such as voice, video, and data traffic. Since exchange carriers have an extensive embedded ATM switching infrastructure, these carriers are using BPON technology to deploy the *fiber-to-the-premises* (FTTP) networks described in Chapters 10 and 12. The impetus behind this is that in addition to being a proven technology, ATM has scalable and flexible traffic-management capabilities and robust quality-of-service features. Chapter 7 has further details on ATM characteristics for BPON use.

6.5.2 EPON and EFM

The widespread use of Ethernet in both local area and metro networks makes it an attractive alternative transport technology for access networks. Since this method encapsulates and transports data in Ethernet frames, it is easy to carry IP packets over an Ethernet link. This scheme thus simplifies the interoperability of metro and wide area network assets with installed Ethernet LANs compared to the use of BPON technology.

The general term for the use of Ethernet in access networks is *Ethernet-in-the-first-mile* (EFM). Three different EFM physical transport schemes are possible. Table 6.2 lists the main characteristics of these three options. They are spelled out in more detail in the IEEE 802.3ah EFM standard, which was approved in June 2004. One EFM option uses an EPON architecture that follows the standard PON layout, which has one main feeder line going to an optical splitter. Up to 32 distribution branches leave the splitter and interface to ONTs. As noted in Table 6.2, the IEEE

TABLE 6.2 Main Physical Layer Characteristics of Three EFM Options

EFM Scheme	Physical Layer Options
EPON	1. 10-km distance; 1 Gbps; 1 × 16 splitter; one bidirectional single-mode fiber
	2. 20-km distance; 1 Gbps; 1 × 16 or 1 × 32 splitter; one bidirectional single-mode fiber
P2P over fiber	1000BASE-LX: extended-temperature-range optics
	1000BASE-X: 10 km over one bidirectional single-mode fiber
	100BASE-X: 10 km over one bidirectional single-mode fiber
P2P over copper	750-m distance; 10 Mbps full-duplex transmission over single-pair non-loaded voice-grade copper wires

802.3ah EFM standard specifies operational conditions for a maximum transmission distance of either 10 or 20 km between the OLT and an ONT.

The other two EFM options employ point-to-point links over either copper wires or optical fibers that directly connect users and the central office. As described in Chapter 8, one alternative for a point-to-point Ethernet optical access network is to run dedicated fibers between a central office and individual subscribers. Such a scenario requires a large number of optical fiber lines, with each line having its own optical transceivers. For example, suppose that the network serves 16 subscribers. If the optical link running to an individual subscriber is bidirectional, this scheme needs 16 fibers. In case the links are unidirectional, a total of 32 optical fibers are needed. Since each subscriber link needs transmitters and receivers at each end, the system needs a total of 32 optical transceivers. Therefore, this type of access network implementation is useful only if each subscriber requires close to the full capacity offered by a gigabit Ethernet line.

6.5.3 GPON Basics

The growing demand for higher speeds in the access network spawned the idea of developing a PON with capabilities beyond those of the BPON and EPON architectures. A major aim of this idea was to develop a versatile PON with a frame format that could transmit variable-length packets efficiently at gigabit per second rates. The FSAN group started such an effort in April 2001. The result was the ITU-T Recommendation series G.984.1 through G.984.4 for a *gigabit PON*.

The network layout for a GPON follows that of a standard PON concept. It also retains much of the same functionality characteristic of BPON and EPON schemes, such as DBA and the use of *operations, administration, and maintenance* (OAM) messages. However, in contrast to BPON and EPON architectures, which were developed from an equipment vendor point of view, the GPON operational scheme is a more customer-driven design. This is reflected in the GPON service requirements specification described in G.984.1. This document takes into account the collective requirements of the leading communication service providers in the world. The GPON details are provided in Chapter 9.

6.6 OPTICS PATH ATTENUATION RANGES

A main concern of the various PON specifications is to prescribe the attenuation ranges that are permitted between the optical transmitter and receiver at transmission distances of 10 and 20 km. For power-budget calculations, the three classes of worst-case attenuation ranges are

- Class A: 5 to 20 dB
- Class B: 10 to 25 dB
- Class C: 15 to 30 dB

Included in these attenuation ranges are losses resulting from the optical fiber, splices, connectors, and optical splitters. In addition, the link designer needs to take into

TABLE 6.3 Longest PON Transmission Distance Allowed Under Ideal Conditions

Optical Splitting Ratio	Splitter Loss (dB)	Distance per Optical Class (km)		
		Class A	Class B	Class C
1 : 16	14	13	33	53
1 : 32	17	2	22	42
1 : 64	20	0	11	31

account the possible occurrence of other link degradations, such as additional splices and fiber lengths resulting from cable repairs, the effects of environmental factors on cable performance, and unforeseen degradations in any passive components.

Table 6.3 shows the longest transmission distance achievable under ideal conditions with high-quality components for the maximum power budgets within the three different optical network classes. The fiber is assumed to have an attenuation of 0.25 dB/km at 1550 nm. Distance calculations at 1310 nm need to take into account a higher fiber attenuation of nominally 0.5 dB/km. For each class the maximum distance is given for optical splitting ratios of 1 : 16, 1 : 32, and 1 : 64. The details of the calculations leading to these numbers are presented in Chapter 11.

6.7 STANDARDS DEVELOPMENT

International standards are essential to communication systems to ensure operational compatibility between the equipment from different manufacturers and to ensure architectural uniformity in a particular PON application. In this section we describe the main organizations that have been responsible for creating PON standards and list some of the key standards for various PON schemes. Details of the standards for BPON, EFM, and GPON implementations are given in Chapters 7 through 9, respectively.

6.7.1 ITU-T

The *International Telecommunication Union* (ITU) is a leading creator and publisher of telecommunication technology, regulatory, and standards information. It is an international organization operating within the United Nations system and is headquartered in Geneva, Switzerland. The ITU provides a forum where governments and private-sector organizations coordinate standards and regulations for global telecom networks and services. The ITU *Telecommunication Standardization Sector* (ITU-T) is one of the three basic sectors of the ITU. The mission of ITU-T is to ensure efficient, on-time production of high-quality standards, usually referred to as *recommendations*, which cover all areas of telecommunications.

6.7.2 FSAN

The major standard used today for access networks was developed by a group of telecommunications companies that formed a policy development group called the

Full Service Access Network (FSAN). In 1995 a group of seven telecommunication service companies formed a committee to develop ATM-based passive optical networks. This group expanded to include many more companies in the following years.

The goal of the FSAN committee was to define an access network architecture that would provide a full spectrum of broadband and narrowband telecommunication services. The challenge in doing this was to take into account the diverse geographies where an access network would be deployed, the wide range of telecommunication regulations in various countries worldwide, different service implementation time frames in these countries, and the use of already installed network infrastructures. In addition, a key factor was to keep installation costs low, since an access network serves a limited number of users. This is in contrast to a long-haul network, where installation costs can be spread over a very large user population. Thus, the importance of mass production of relatively inexpensive equipment and components was an important consideration.

Given these requirements, the FSAN committee recognized that the use of an optical splitter in a passive optical network and the transport of ATM cells over this network should be two key drivers for low installation costs and optimum use of a large base of installed ATM equipment. After producing a set of technical specifications, the FSAN committee submitted them to the ITU-T for consideration. In October 1998 their efforts resulted in the approved international standard known as ITU-T Recommendation G.983.1 for an ATM PON.

Subsequent activity of the FSAN committee included defining requirements for BPON and GPON architectures. As was the case for the APON initiative, the committee brings their recommendations to the ITU-T, which then develops agreed-upon standards, such as the G.983 BPON series, the G.984 GPON family, and Recommendation G.985 for an EPON.

6.7.3 IEEE

The *Institute of Electrical and Electronics Engineers* (IEEE) is an international organization headquartered in the United States that is involved with all aspects of electrical and electronic engineering. One aspect of the IEEE is to develop standards involving an area of interest within the scope of the IEEE. For example, the IEEE 802 Committee develops LAN and MAN standards, two popular ones being the extensive families of IEEE 802.3 Ethernet and the IEEE 802.11 wireless standards.

In relation to access networks, in November 2000 the IEEE 802.3 Ethernet Working Group formed an *Ethernet in the First Mile Study Group*. The main objectives were to work on point-to-point Ethernet on optical fiber, Ethernet passive optical networks on point-to-multipoint networks, and a physical layer specification for using point-to-point copper wires in access links. In July 2001, the IEEE approved the IEEE 802.3ah EFM standard recommended by the study group.

6.8 SUMMARY

Given that network and service providers are seeking to reduce their operational costs, the concept of using a passive optical network (PON) is an attractive option.

In a PON there are no active components between the central office and the customer's premises. Instead, completely passive optical components are placed in the network transmission path to guide the traffic signals contained within specific optical wavelengths to the user endpoints.

In the basic architecture of a typical PON, a fiber optic network connects switching equipment in a central office with a number of service subscribers. Starting at the central office, one single-mode optical fiber strand runs to a passive optical power splitter near a housing complex, an office park, or some other campus environment. The splitting device simply divides the optical power into separate paths to the subscribers. The number of splitting paths can vary from 2 to 64, but typically they are 8, 16, or 32. From the optical splitter, individual single-mode fibers then run to each building or serving equipment. The transmission span from the central office to the user can be up to 20 km.

Simultaneous transmission of separate service types on the same fiber is enabled by using different wavelengths for each direction. Downstream transmissions use a 1490-nm wavelength for voice and data and a 1550-nm wavelength for video distribution. Upstream voice and data traffic use a 1310-nm wavelength. Depending on the particular PON standard being used, the downstream and upstream transmission equipment operate at 155 Mbps, 622 Mbps, 1.25 Gbps, or 2.5 Gbps. In some cases the transmission rates are the same in either direction (a symmetric network). In other PON standards the downstream rate may be higher than the upstream rate, which is called an asymmetric implementation.

Time-division multiplexing (TDM) is the basic protocol used to combine incoming voice and data streams that are destined for users on the PON. The multiplexed downstream signal is broadcast to all the subscribers. Each subscriber discards or accepts the incoming information packets, depending on the packet header addressing. To avoid collisions between the transmissions of different users in the upstream direction, the system uses a time-division multiple access (TDMA) protocol.

The three main PON alternatives are known as broadband PON (BPON), Ethernet PON (EPON), and gigabit PON (GPON). An architecture consisting of point-to-point Ethernet links is a competing access network technology. Ethernet use in access networks is referred to as Ethernet-in-the-first-mile (EFM). Details on BPON, EPON, and GPON are given in Chapters 7 through 9, respectively.

PROBLEMS

6.1 What are some advantages and drawbacks to implementing a passive optical network by using a configuration with ONUs instead of ONTs?

6.2 Compare the following two network architectures: (1) a PON based on using ONUs as the termination points; (2) a hybrid fiber–coaxial network (see Section 1.2).

6.3 Using vendor data sheets (e.g., as can be found on the Internet) for PON or FTTP optical network terminations (ONTs), list some of the telecommunication service interfaces (e.g., Ethernet, T1/E1, ATM) of several ONTs with various levels of capabilities.

6.4 (a) Why may it be more important to encrypt information in the downstream direction in a PON and not necessarily in the upstream path?

(b) How might information be compromised either maliciously or unintentionally in either direction?

(c) Find out how encryption might be implemented in the upstream and downstream directions.

6.5 In terms of network configuration flexibility and ease of maintenance and troubleshooting, at what cable interfaces to units such as optical splitters, fiber termination boxes, patch panels, and WDM couplers would it be better to use optical connectors instead of splices?

6.6 (a) Find some commercially available thin-film filters (TFFs) that can be used for the WDM coupler setup shown in Figure 6.9.

(b) Find some commercially available TFFs that can be deployed in the setup shown in Figure 6.10.

6.7 Assume that an optical signal experiences the following losses in an ideal PON link: 0.6 dB due to connectors and splices, fiber attenuation of 0.27 dB/km, and the splitter loss shown in Table 6.3. If the network designer includes a 2.0-dB safety margin to account for any unforeseen losses, verify the maximum transmission distances shown in Table 6.3 for the three splitter types using class B optics.

6.8 Based on information that can be found on their Web sites, describe the available standards and ongoing standardization initiatives related to PON and FTTP implementations of the ITU-T and the IEEE.

6.9 Make a detailed comparison chart of BPON, EPON, and GPON concepts based on both technology and standards points of view. Why was each particular methodology conceived?

FURTHER READING

1. BiCSI, *Telecommunications Cabling Installation*, McGraw-Hill, New York, 2001.
2. A. B. Carlson, *Communication Systems*, 4th ed., McGraw-Hill, Burr Ridge, IL, 2002.
3. B. Chomycz, *Fiber Optic Installer's Field Manual*, McGraw-Hill, New York, 2000.
4. S. Devadhar and K. Ryan, "Dynamic bandwidth allocation over passive optical networks," *Lightwave*, vol. 17, pp. 138–142, Nov. 2000.
5. B. A. Forouzan, *Introduction to Data Communications and Networking*, 2nd ed., McGraw-Hill, Burr Ridge, IL, 2001.
6. P. E. Green, "Fiber to the home: the next big broadband thing," *IEEE Commun. Mag.*, vol. 42, pp. 100–106, Sept. 2004.

7

BPON CHARACTERISTICS

Broadband passive optical network (BPON) standards are based on the G.983 series of ITU-T Recommendations that specify *asynchronous transfer mode* (ATM) as the transport and signaling protocol. Since telecommunication services carriers have an extensive embedded ATM switching infrastructure, these carriers are using BPON technology to deploy the *fiber-to-the-premises* (FTTP) networks described in Chapters 10 and 12. The impetus behind this is that in addition to being a proven technology, ATM has scalable and flexible traffic-management capabilities and robust quality of service features.

Section 7.1 covers the fundamental BPON architecture, which includes factors such as the network layout, downstream and upstream data rates, and assignments of operational wavelengths. Since ATM is the basis of the BPON specifications, in Section 7.2 we give an overview of ATM concepts and describe what types of service levels users can subscribe to. The BPON operational concepts, including bidirectional traffic flows, packet encapsulation, and operations, administration, and maintenance functions, are presented in Section 7.3. Since a BPON uses a shared transmission medium, an essential BPON function is to have orderly and efficient traffic control. In Section 7.4 we discuss methodologies for this control. Finally, in Section 7.5 we outline the constituents of the G.983 series of ITU-T Recommendations for BPON implementations. These BPON standards include quality-of-service guidelines based on the ATM transport protocol for various types of services, and they give procedures for the detection, registration, and ranging of installed ONTs.

7.1 BPON ARCHITECTURE

Figure 7.1 shows the basic BPON architecture and operational concept. This architecture follows the standard PON layout described in Chapter 6 with a maximum transmission distance of 20 km between the OLT and an ONT (or an ONU).

FTTX Concepts and Applications, by Gerd Keiser
Copyright © 2006 John Wiley & Sons, Inc.

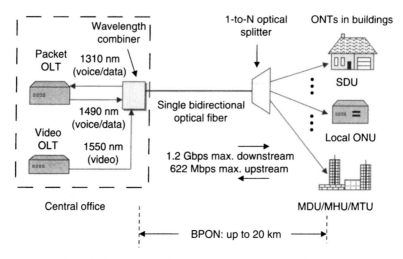

Figure 7.1. Basic BPON architecture and operational concept.

7.1.1 Traffic Flow Schemes

Downstream transmission uses a 1490-nm wavelength for combined voice and data traffic, which gets encapsulated in ATM cells by the switching equipment at the OLT. As described in Section 7.2, ATM cells are 53-byte-long packets that contain 48 bytes of information and have a 5-byte control header. The BPON G.983 standard offers a maximum bandwidth of 1.2 Gbps for the downstream traffic. Combined voice and data traffic is transmitted upstream by the ONT using a 1310-nm wavelength. This upstream wavelength uses the same fiber that carries the downstream traffic. The BPON standard allows a data rate of up to 622 Mbps for upstream traffic.

The combining and separating of the three independent wavelengths at the OLT and at the ONT are done using the types of CWDM couplers described in Section 6.4. In an actual FTTP implementation, these couplers could be located in the same equipment rack in which the OLT equipment resides or they could be in a *fiber distribution frame* (FDF), which serves as the termination point of the fibers in the feeder cable coming into the central office from the outside cable plant. In Section 12.1 we discuss the criteria for selecting a particular location for these wavelength couplers.

A separate 1550-nm wavelength is used for transmitting video traffic downstream from the OLT. A major application for this one-way traffic is *video distribution*. This application enables a viewer to use a set top box to select any of a large number of TV channels. The video traffic is sent independent of the ATM-encapsulated voice and data traffic. The video signals can be in either analog or digital format, and both formats can be sent simultaneously as independent video channels in different frequency bands. In addition, both standard television and high-definition television (HDTV) channels can be sent simultaneously over the same fiber at the 1550-nm wavelength. In Appendix C we provide further details on video transmission requirements.

Downstream 1490-nm voice and data traffic is transmitted using *time-division multiplexing* (TDM) of the ATM cells. Upstream 1310-nm voice and data traffic (in the form of ATM cells) is transmitted by means of a *time-division multiple access* (TDMA) protocol. In Section 7.3 we give the details of these protocols for a BPON.

7.1.2 OLT Capabilities

In an actual network, a single OLT can manage several separate PONs simultaneously. Figure 7.2 illustrates this for one OLT controlling 22 BPONs. Table 7.1 lists specifications for such an OLT shelf. Note that the specifications for the particular OLT unit described in this table are for illustration purposes only. The reader should refer to current vendor product data sheets for the latest BPON equipment specifications.

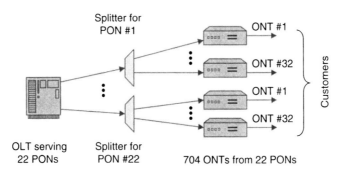

Figure 7.2. Single OLT that manages 22 separate BPONs simultaneously.

TABLE 7.1 Specifications for a Triple-Play OLT That Can Serve 22 PON Lines

Item	Characteristic
Transmission method	FSAN BPON Standard (ITU-T G.983 family)
Transmission speed	Downstream: 622 Mbps; upstream: 155 Mbps
Wavelengths	Upstream: 1310 nm voice/data; downstream: 1490 nm voice/data, 1550 nm video
Transmission distance	20 km maximum
Number of PON lines	22 maximum (1 per PON interface card)
Number of branches (users)	32 maximum per line (704 per OLT)
Network interface	100BaseTX
Power consumption	380 W approximate
External dimensions	$W \times D \times H$: 465 mm × 457 mm × 444 mm (18.3 in. × 18 in. × 17.5 in.)
Shelves per rack	Four shelves per 7-ft rack 19- or 23-in. rack mountable
Weight	Approx. 30 kg (66 lb.)

An OLT that is capable of controlling multiple BPONs can be used in FTTP applications for services to residences, multiple dwelling units (MDUs), *small office/home office* (SOHO) users, and small to medium-sized enterprises. Up to 32 end users can share each BPON optical fiber line connecting to the OLT. Thus, given that there are 22 BPON lines per OLT, it can deliver triple-play services to over 700 customers. The size of a standard OLT shelf allows four units to fit into one 7-ft (2-m) rack, thereby enabling voice, video, and data service delivery to 2816 customers per equipment rack. In addition, an OLT typically incorporates the highest levels of carrier-class redundancy specified for protection switching in SONET/SDH equipment in the event of line or circuit card failures.

7.2 ATM BASICS

Asynchronous transfer mode is a high-performance switching and multiplexing technology that utilizes fixed-length packets to carry different types of traffic. A key attraction of ATM is that it enables carriers to offer multiple classes of service over wide area networks (WANs), to connect devices that operate at different speeds, and to mix a variety of traffic types having different transmission requirements, such as voice, video, and data traffic.

7.2.1 Use of ATM Cells

The ATM scheme formats all information into fixed-length packets (called *cells*), which consist of 48 bytes of *information payload* and 5 bytes of *overhead*, as Figure 7.3 illustrates. The function of the 5-byte header is to enable efficient high-speed switching. It contains payload-type information, virtual-circuit identifiers, and a header error-check function.

The advantage of using a fixed cell size is to ensure that time-sensitive information such as voice or video services arrives at the destination in a timely fashion. Since voice and video packets normally are small compared to data packets, mixing the two types of traffic using ordinary multiplexing techniques can cause unacceptable delays in the time-sensitive traffic. Figure 7.4 gives an example of this for a stream of short voice packets (1, 2, 3, etc.) being multiplexed with a string of long data packets (A, B, etc.). If standard time-division multiplexing occurs as shown, there is a long delay between voice packets 1 and 2, for example, since the long data packet A is sandwiched between them.

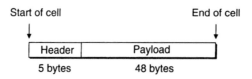

Figure 7.3. An ATM cell consists of 48 bytes of payload and 5 bytes of overhead.

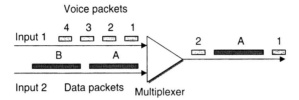

Figure 7.4. Multiplexing packets of different sizes can cause delay in small packets.

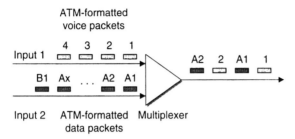

Figure 7.5. Multiplexing of ATM cells mitigates delays caused by long packets.

When using ATM, both the voice and the data packets are divided into equal-length cells, as illustrated in Figure 7.5. Now when the cells from the voice and data streams are interleaved at the multiplexer output, neither cell stream suffers a long delay. Furthermore, since ATM uses high-speed transmission links, the small 53-byte cells arrive at their respective destinations in the same continuous stream as they arrived at the multiplexer, despite having been interleaved with cells from other traffic streams. The ATM equipment at the destination reassembles the cell payloads into the original data packet format. This means that an ATM network can carry both time-sensitive traffic, such as real-time video, and non-time-critical data traffic without any of the users being aware of the underlying packet segmentation and reassembly processes.

7.2.2 ATM Service Categories

A key characteristic of ATM is that it can transfer a diverse set of traffic types simultaneously, including both real-time and non-real-time traffic. *Real-time services* need to deliver information in a timely fashion and as a continuous, smooth flow. Examples of this are telephone conversations, video conferencing, television, and distance learning. Any delay greater than a few tenths of a second or any loss-induced gaps in the information flow will cause a significant degradation in the quality of service. *Non-real-time services* are used for applications that do not have tight constraints on delay or delay variations. Examples of this are data and still-image transfers, airline reservations, and banking or investment transactions.

The ATM protocol has a set of *quality-of-service* (QoS) attributes that are associated with the performance of a network connection. Some of these attributes are

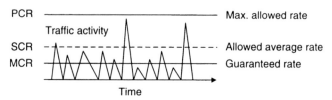

Figure 7.6. PCR, SCR, and MCR.

user-oriented, whereas others are related to the network. The *user-oriented attributes* define the rate at which a user desires to send data. The constraints on these rates are established in a service contract and are verified between a user and a network at the time of connection establishment. They include the following basic parameters, which Figure 7.6 illustrates:

1. *Peak cell rate* (PCR): defines the maximum rate in cells per second at which a user is allowed to transmit according to the service contract rules.
2. *Sustainable cell rate* (SCR): the average allowed cell rate measured in cells per second over a long time period. The actual cell rate may be higher or lower than this value during any time interval.
3. *Minimum cell rate* (MCR): the transmission rate guaranteed to a user by the service provider regardless of the state of network congestion.

In many networks the traffic rate heading upstream is much lower than the rate of the downstream traffic going to the user. For example, a user might send only small amounts of data upstream, such as Internet search requests or e-mail messages. However, the search requests can result in large files being sent back to the user at a high rate. For such network conditions, either the customer can request different speed capabilities for upstream and downstream data, or in many cases an Internet service provider already has these features built into the service offering. This arrangement results in higher network efficiency and lower operating costs.

The *network-related attributes* define the characteristics of the network. The key parameters include the following:

1. *Cell loss ratio* (CLR): the percentage of cells not delivered to their destination. These cells could be lost or delivered so late that they are considered lost. The loss in the network could be due to congestion or buffer overflow.
2. *Cell error rate* (CER): designates the fraction of cells that are delivered with errors.
3. *Cell transfer delay* (CTD): the delay experienced by a cell between the time it entered the network at the source and the time it exited at the destination. It includes propagation delays, queuing delays at various intermediate switches, and service times at queuing points.
4. *Cell delay variation* (CDV): measures the time difference between the maximum and minimum CTD.

TABLE 7.2 Summary of Descriptions and Uses of the Five ATM Classes of Service

Service Class	Quality-of-Service Parameter	Example Applications
Constant bit rate (CBR)	The CBR cell rate is constant with time to satisfy applications that are sensitive to cell-delay variations.	Telephone traffic, television, video-on-demand, and video conferencing.
Variable bit rate Non-real time (VBR-nrt)	VBR allows users to send traffic at a rate that may vary with time; that is, the traffic rate depends on the availability of user information. Statistical multiplexing may be provided to make optimum use of network resources.	Airline reservations and banking or investment transactions.
Real time (VBR-rt)	This service class is similar to VBR-nrt but is intended for applications that are sensitive to cell-delay variations.	Voice with speech-activity detection and interactive compressed video.
Available bit rate (ABR)	ABR provides rate-based flow control. Depending on the degree of network congestion, the source is required to control its rate. Users may declare a minimum cell rate, which is guaranteed by the network.	File transfers and e-mail.
Unspecified bit rate (UBR)	Cells are forwarded on a first-in, first-out basis using the capacity not consumed by other services. Both delays and variable losses of cells are possible.	Transfer, messaging, distribution, or retrieval of text, data, or images.

To enable the simultaneous transfer of diverse traffic types, the ATM standard defines five classes or categories of service. Two of these relate to real-time services and the other three to non-real-time services. Table 7.2 summarizes the service classes, the associated QoS parameters, and example applications.

The ATM service classes have the following characteristics:

1. *Constant bit rate* (CBR). CBR provides the highest-priority service. This service class is designed for ATM subscribers who need a fixed amount of bandwidth that is available continuously for the duration of the active connection. When a CBR connection is established, the user negotiates a guaranteed PCR with the network. The user then can send ATM cells at any rate up to the PCR at any time and for any duration. CBR supports real-time applications such as voice, videoconferencing and video distribution, interactive multimedia, and circuit emulation.

2. *Real-time variable bit rate* (rt-VBR). VBR is used for bursty traffic sources, which require small delay variations (jitter). When establishing an rt-VBR connection, the user negotiates the PCR, an SCR value, and a *maximum burst size* (MBS) with the network. In addition, a maximum cell delay variation time is specified. The SCR can be viewed as an average cell rate over time. If the user traffic exceeds the negotiated PCR, cells submitted in excess of the PCR are dropped. This service class is intended for real-time applications such as packet-formatted voice, real-time and near-real-time video, and time-sensitive data.

3. *Non-real-time variable bit rate* (nrt-VBR). To set up a nrt-VBR connection, a user negotiates the same PCR, SCR, MBS, and CDV parameter values as for a rt-VBR connection. The only difference is that in the nrt-VBR case the network does not guarantee the cell delay performance. Typical applications include frame relay service, store-and-forward non-real-time video and audio, and client–server applications such as banking transactions and reservation systems.

4. *Available bit rate* (ABR). For an ABR connection the user negotiates a PCR value and a *minimum cell rate* (MCR). The network then guarantees the delivery of cells sent at or below the MCR. Cells submitted in excess of the PCR are dropped. ABR is suited for applications such as credit card validation, fax, e-mail, Internet access, and general file transfers.

5. *Unspecified bit rate* (UBR). This is a best-effort cell delivery service. For a UBR connection the user negotiates only a PCR value. There are no other performance guarantees. Cells submitted in excess of the PCR are dropped. This means that if the network is congested, there will probably be a high level of cell loss. UBR typically is used for the same types of applications as ABR.

7.2.3 Service Level Agreements

A *service level agreement* (SLA) is a contract between a service provider and its customers. The purpose of an SLA is to define and guarantee an acceptable level of network performance. This can include factors such as response time, availability of service, downtime, security characteristics, and repair response time. Typically, if the service provider does not meet the SLA agreements, the customer will receive some type of credit for the lost service time.

7.3 BPON OPERATIONAL CHARACTERISTICS

Important operational aspects for maintaining a high quality of service in a BPON include efficient bandwidth utilization, management of buffer content at the ONTs, and proper synchronization of upstream traffic over the shared PON medium. In Section 7.4 we describe bandwidth utilization issues together with buffer management. In this section we first address traffic flow issues for the voice and data channels and then describe some options for the separate video channel.

7.3.1 Voice and Data Traffic Flows

The OLT in the central office is the interface between the exchange carrier network and the subscribers on the BPON. When information content in the form of voice or data arrives at the OLT from the carrier network, the OLT sends the information in the form of ATM cells to all ONTs on the BPON using a time-division multiplexing (TDM) scheme. The downstream BPON bit rate options are 155.52, 622.08, and 1244.16 Mbps.

A special frame and time-slot structure is used to send and receive cells. As shown in Figure 7.7 for a 155.52-Mbps rate, the downstream frame format consists of 56 cells that have the 53-byte ATM cell length. There are two types of downstream cells. *Data cells* carry information including user data, signaling information, and ATM *operation, administration, and management* (OAM) information.

The second type are *physical layer OAM* (PLOAM) cells. These cells are responsible for synchronization, error control, security, maintenance, and bandwidth allocation. A 155.52-Mbps downstream frame contains two PLOAM cells, one at the beginning of the frame and the other in the middle, and 54 data cells. Since only 54 of the 56 cells carry information, the effective downstream rate is

$$54/56 \times 155.52 \text{ Mbps} = 149.97 \text{ Mbps}$$

At a 155.52-Mbps transmission rate the time length of the frame is 152.67 µs. The downstream frame size scales up to 224 cells for a 622.08-Mbps rate and to 448 for a 1244.16-Mbps rate. These two higher rates contain eight and 16 PLOAM cells, respectively.

Figure 7.8 shows the upstream frame format for a 155.52-Mbps rate. In this case there are 53 cells, each of which is 56 bytes long. The additional 3 bytes for each cell are overhead, which the OLT can program for various functions. The three overhead bytes contain a minimum of 4 bits of guard time, a preamble, and a delimiter field. The *guard time* provides a sufficient distance in time between cells to prevent collisions with cells coming from other ONTs. The information in the *preamble field* is used to extract the phase of the incoming ATM cell relative to the local clock in the OLT and may be used to acquire bit synchronization. The *delimiter field* is a unique bit pattern that indicates the start of an incoming cell and can be used to perform byte synchronization. The upstream cell traffic also contains PLOAM cells coming from each ONT. The OLT defines the PLOAM cell rate for each ONT. The minimum rate is one PLOAM cell every 100 ms.

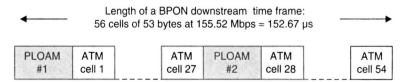

Figure 7.7. Downstream BPON frame format for a 155.52-Mbps rate.

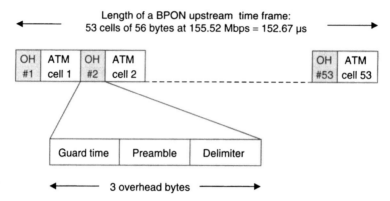

Figure 7.8. Upstream BPON frame format for a 155.52-Mbps rate.

Since the ONTs use *time-division multiple access* (TDMA) to send information to the OLT, each ONT must be synchronized with all the other ONTs. To achieve this the OLT uses a *ranging* process, which determines how far away each ONT is from the OLT. Once this distance is known, the OLT assigns an optimal synchronized time slot in which an ONT can transmit without interfering with other ONTs. This is done through the inclusion of 1-byte *transmission permits* (*grants*) contained within the downstream PLOAM cells. Each PLOAM cell has 27 grants that any ONT can read. However, a frame needs only 53 grants, which are mapped into the two PLOAM cells of a 155-Mbps downstream frame. As a result, the last grant of the second POAM cell is an idle (dummy) grant. In an asymmetric transmission case (e.g., 622 Mbps downstream and 155 Mbps upstream), the grant fields of PLOAM cells 3 through 8 are filled with idle grants that an ONT will not use. When an ONT needs to send information, it checks the data grant number in the first two PLOAM cells and compares it to its own number. If there is a match, the ONT can send its information.

7.3.2 Protection of Grants

A *cyclic redundancy check* (CRC) method is used to protect the integrity of a group of seven grants. As we describe in more detail in Appendix D, a CRC is a polynomial code that is based on a binomial division process. The coding process uses a block of information bits and a sequence of redundant bits to check for errors in the information block. For a downstream BPON link, a CRC protects a group of seven transmission grants in a PLOAM cell. The CRC generating polynomial this group is

$$g(x) = x^8 + x^2 + x + 1$$

Such a polynomial can protect up to 15 bytes and is able to detect up to three bit errors. In the second PLOAM cell the dummy seventh grant which is added to the

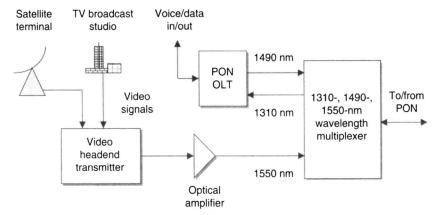

Figure 7.9. Video overlay service to BPON subscribers.

six real grants is used as part of the CRC calculation for that group. No error correction (only error detection) is done in a BPON system. Therefore, when the receiver gets a CRC that is incorrect, the entire block of information bits is ignored.

7.3.3 Video Traffic

As Figure 7.9 shows, the 1550-nm downstream wavelength provides a video overlay service to network subscribers. The video-related equipment in the central office consists of a headend transmitter that receives content from various analog and digital video sources. The transmission system broadcasts these video services to subscribers by means of the same standard subcarrier multiplexed modulation scheme as CATV uses.

Since the overlay uses a separate 1550-nm wavelength, it is transparent to the other voice and data services sent over the same fiber at 1490 nm. Therefore, any coding scheme can be used for the video. These include amplitude-modulated vestigial sideband (AM-VSB), 64- or 256-point quadrature amplitude modulation (QAM64/256), quadrature phase-shift keying (QPSK), or direct MPEG modulation (see Appendix C for further information on these modulation formats).

Immediately following the video transmitter is an optical amplifier that boosts the optical signal level before it is sent out over the BPON feeder cable. This amplification is necessary in order to have a sufficient optical signal-to-noise ratio for the video signal at the customer's premises.

7.4 TRAFFIC CONTROL

Subscribers on an ONT have a wide range of service needs and bandwidth use. As a result, service providers sell individual subscribers a service contract based on their network usage needs. For example, the granularity of service may be offered in $N \times 64$-kbps increments. In early PON implementations the bandwidth allocated to a specific user was guaranteed at a fixed value. This method results in a large amount of bandwidth remaining unused, since subscribers often have no information to send in their allocated time slots. A more efficient method is to allocate bandwidth to each user dynamically, depending on their specific needs at any given moment.

7.4.1 Fixed Bandwidth Allocation

In the simplest bandwidth allocation scenario, the service increment sold to each specific customer is fixed. This *fixed bandwidth allocation* is known popularly as *nailed-up bandwidth*. However, this method is not an efficient use of bandwidth. If a particular subscriber has no information to send during some time interval, that unused bandwidth segment on the network is not available to other subscribers, who may have a large amount of information waiting to be sent.

Figure 7.10 illustrates this scenario. Consider the case where each of three particular ONTs on a PON is allocated 5 Mbps of fixed bandwidth, that is, each has a 5-Mbps CBR service. Now suppose that ONT2 has a large amount of data waiting in a queue but ONT3 does not have any information to send. In this situation only a small fraction of the ONT3 bandwidth is being used to send idle cells in order to keep the link in synchronization. However, since the bandwidth per ONT is fixed, ONT2 cannot use the idle assets of ONT3 to move traffic that is waiting in the queue more quickly. Since data typically arrive in bursts, a large portion of the BPON bandwidth can remain unused with fixed bandwidth assignments.

7.4.2 Dynamic Bandwidth Allocation

When using a fixed bandwidth allocation, buffers start to fill up at an ONT if information that needs to be sent arrives in bursts. This condition results in small delay variations or jitter in cell traffic, which can degrade the signal quality. In addition, when the buffers are full, the resulting traffic congestion can cause

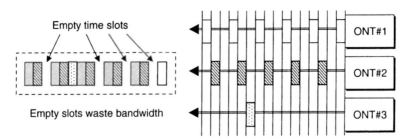

Figure 7.10. Fixed-bandwidth allocation scenario.

cell discards. Minimizing the effects of jitter and traffic congestion to maintain a high quality of service in a BPON can be achieved by allocating bandwidth dynamically according to the changing transmission requirements of individual ONTs.

Dynamic bandwidth allocation (DBA) is a methodology that allows quick reapportioning of bandwidth on the PON based on current traffic requirements. The DBA process is controlled by the OLT, which issues *grants* or permits that allow an ONT to transmit within a specific time slot. The upstream traffic control is done by apportioning these grants to individual *traffic containers* (T-CONTs), which are upstream traffic flow categories defined within an ONT.

To determine how many grants (i.e., how much bandwidth) to assign to an ONT, the OLT needs to know the traffic status of the T-CONT associated with that ONT. Two possible schemes for doing this are the *status reporting* method and the *idle-cell monitoring* procedure. In the status reporting method, as part of its traffic status a T-CONT indicates how many packets are waiting in its buffer. Once the OLT receives this information, it can reapportion the grants to various ONTs accordingly. In contrast, the idle-cell monitoring procedure is done only at the OLT. When an ONT has no information waiting to be transported, upon receiving a grant it sends an idle cell upstream to indicate that its buffer is empty. This informs the OLT that the grants for that T-CONT can be assigned to other T-CONTs.

If an ONT has a long queue waiting in its buffer, the OLT can assign multiple T-CONTs to that ONT. In addition, the T-CONTs can be characterized into five different types, depending on the operational parameters spelled out in the SLA for a particular ONT, such as service priority and QoS. The service priority is given in terms of *assignable bandwidth*, ranging from high to low priority, as Figure 7.11 illustrates. The *fixed bandwidth* is guaranteed and is not dynamically controllable. However, since it is fixed at a constant value whether or not communication is taking place, it has the highest priority. *Assured bandwidth* is also guaranteed, but is assigned dynamically only during actual communications. Thus, it also has a high priority. The *nonassured bandwidth* and *best-effort bandwidth* types are assigned dynamically under the control of the DBA function. Table 7.3 summarizes the T-CONT types and the associated operational parameters. For more details, the interested reader can consult the G.983.4 ITU-T Recommendation.

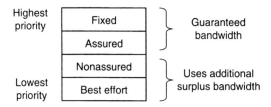

Figure 7.11. Relationship between priority and assignable bandwidth.

TABLE 7.3 T-CONT Definitions and Corresponding Services

Traffic Container	Assignable Bandwidth	Characteristics
T-CONT 1	Fixed	Guaranteed bandwidth for time-sensitive applications; specified in G.983.1
T-CONT 2	Assured	Guaranteed bandwidth for non-time-sensitive applications
T-CONT 3	Nonassured	Minimum guaranteed bandwidth plus additional surplus nonguaranteed bandwidth for non-time-sensitive applications
T-CONT 4	Best effort	Dynamically assigned surplus bandwidth independent of the assured bandwidth
T-CONT 5	All	Mixture of all service categories

7.5 STANDARDS DETAILS

The first BPON standard was the ATM-based ITU-T Recommendation G.983.1, which was ratified in 1998. This was referred to as an *APON standard*. The APON transmission rates were specified as 622.08 Mbps for downstream traffic and 155.52 Mbps for upstream packet traffic. All services (voice and data) sent over an APON are encapsulated in data packets. An APON uses a separate coax cable or fiber for RF video distribution.

To improve on this APON scheme, the ITU-T developed the G.983 series of recommendations. The initial enhancements were first to add another wavelength for video services on the same fiber and then to increase the downstream transmission rate to 1.244 Gbps and the upstream rate to 622 Mbps. These are included in the BPON Recommendation G.983.3, which the ITU-T ratified in December 2001. In the following paragraphs we give an overview of the material described in G.983.1 through G.983.8.

7.5.1 Recommendation G.983.1

Recommendation G.983.1 was approved in October 1998 and is called *Broadband optical access systems based on passive optical networks (PON)*. It describes a flexible optical fiber access network that can support the bandwidth requirements of ISDN (Integrated Services Digital Network) and B-ISDN (broadband ISDN) services. The recommendation addresses systems with nominal symmetrical line rates of 155.520 Mbps and asymmetrical line rates of 155.520 Mbps upstream and 622.080 Mbps downstream. It also proposes specifications for the *physical media-dependent* (PMD) layer, the *transmission control* (TC) layer, the process for detection and registration of an installed ONT, and the ONT ranging protocol.

7.5.2 Recommendation G.983.2

Recommendation G.983.2 was approved in June 2002 and is called *ONT management and control interface specification for B-PON*. It gives requirements for managing

ONTs through the OLT using the *ONT management and control interface* (OMCI). First, it specifies managed entities of a protocol-independent *management information base* (MIB) that describes the information exchange procedure between the OLT and ONT. Then it covers the ONT management and control channel, its protocol, and details of management and control messages.

7.5.3 Recommendation G.983.3

Recommendation G.983.3 was approved in May 2001 and is called *A broadband optical access system with increased service capability by wavelength allocation*. It defines new wavelength allocations to distribute ATM-PON signals and additional service signals (such as video) simultaneously. Potential applications include video broadcast distribution or data services, which could be sent unidirectionally or bidirectionally. Detailed specifications of these services, such as modulation scheme and signal format, are not covered.

7.5.4 Recommendation G.983.4

Recommendation G.983.4 was approved in November 2001 and is called *A broadband optical access system with increased service capability using dynamic bandwidth assignment*. It specifies requirements for adding dynamic bandwidth assignment (DBA) functionality to the systems defined in ITU-T Rec. G.983.1. The recommendation gives specifications for DBA operation and for DBA-related communication between the OLT and the ONUs or ONTs.

7.5.5 Recommendation G.983.5

Recommendation G.983.5 was approved in January 2002 and is called *A broadband optical access system with enhanced survivability*. It describes the functions needed to extend ITU-T Rec. G.983.1 to enable survivability and network protection enhancements for delivering highly reliable services. The document includes BPON survivability architectures, protection performance criteria, and protection-switching criteria and protocols.

7.5.6 Recommendation G.983.6

Recommendation G.983.6 was approved in June 2002 and is called *ONT management and control interface specifications for B-PON system with protection features*. The document describes the additional OMCI specifications that are required to support the protected BPON systems described in ITU-T Recommendation G.983.5. The enhancements include the addition of protection-related attributes to existing managed entities, managed entity relations diagrams for protected systems, detailed scenarios for OMCI functions (start up, tear down, and switch over), and methods for the configuration or removal of virtual paths (VP) in protected BPON systems.

7.5.7 Recommendation G.983.7

Recommendation G.983.7 was approved in November 2001 and is called *ONT management and control interface specification for dynamic bandwidth assignment (DBA) B-PON system*. The document describes the OMCI specifications required for the DBA function in a BPON. The recommendation enhances the core functions specified in G.983.2 and specifies managed entities of a protocol-independent MIB that models the exchange of information between the OLT and an ONT.

7.5.8 Recommendation G.983.8

Recommendation G.983.8 was approved in March 2003 and is called *BPON OMCI support for IP, ISDN, video, VLAN tagging, VC cross-connections, and other select functions*. This recommendation provides OMCI support for the BPON system defined in G.983.1 for select functions that were not included in G.983.2. OMCI support is specified for Internet protocol (IP) router functionality on local area network cards, ISDN interfaces, additional Ethernet performance monitoring, video interfaces, virtual LAN (VLAN) tagging, extended media access control (MAC) bridge filtering, local craft terminal interfaces, and virtual channel (VC) cross-connections.

7.6 SUMMARY

Broadband passive optical network (BPON) standards are based on the G.983 series of ITU-T Recommendations that specify asynchronous transfer mode (ATM) as the transport and signaling protocol. Since telecommunication services carriers have an extensive embedded ATM switching infrastructure, these carriers are using BPON technology to deploy fiber-to-the-premises (FTTP) networks. The impetus behind this is that in addition to being a proven technology, ATM has scalable and flexible traffic-management capabilities and robust quality-of-service features.

Downstream transmission uses a 1490-nm wavelength for combined voice and data traffic, which gets encapsulated in ATM cells at the OLT. The BPON G.983 standard offers a maximum bandwidth of 1.2 Gbps for the downstream traffic. Voice and data traffic is transmitted upstream by the ONT using a 1310-nm wavelength at a data rate of up to 622 Mbps.

A separate 1550-nm wavelength is used for transmitting video traffic downstream from the OLT. A major application for this one-way traffic is video distribution. The video signals can be in either analog or digital format, and both formats can be sent simultaneously as independent video channels in different frequency bands. In addition, both standard television and high-definition television (HDTV) channels can be sent simultaneously over the same fiber at the 1550-nm wavelength.

Downstream 1490-nm voice and data traffic is transmitted using time-division multiplexing (TDM) of the ATM cells. Upstream 1310-nm voice and data traffic (in the form of ATM cells) is transmitted by means of a time-division multiple access (TDMA) protocol.

Dynamic bandwidth allocation (DBA) is an efficient dynamic methodology that allows quick reapportion of bandwidth on the PON based on current traffic requirements. The DBA process is controlled by the OLT, which issues *grants* or permits that allow an ONT to transmit within a specific time slot. The upstream traffic control is done by apportioning these grants to individual traffic containers (T-CONTs), which are upstream traffic flow categories defined within an ONT.

PROBLEMS

7.1 Using vendor data sheets (e.g., as can be found on the Internet) for PON or FTTP optical line terminations (OLTs), list some of the telecommunication service interfaces (e.g., Ethernet, T1/E1, ATM, video) of several OLTs with various levels of capabilities. How many independent PON lines can be served from an individual OLT from various vendors?

7.2 Consider an ATM network in which 8 out of 10,000 cells are lost during transmission and 4 out of 10,000 contain errors.
  (a) What is the cell loss ratio (CLR)?
  (b) What is the cell error ratio (CER)?

7.3 To what ATM service classes would the following applications belong?
  (a) Downloading a document from the Internet
  (b) A telephone call between people in two cities
  (c) Online purchase of stock
  (d) Medical consultation by means of a telemedicine application
  (e) Sending an e-mail message

7.4 Consider a 1.544-Mbps link that uses 5 percent of the bits for framing purposes. How many 64-kbps signals can be transported over this link using an ATM protocol that maps 47-byte segments of information plus a 1-byte header into the 48-byte ATM cell payload?

7.5 A user needs an ATM service guarantee to send information at a rate of 256 kbps using a CBR protocol that maps 47-byte segments of information plus a 1-byte header into the 48-byte ATM cell payload. Occasionally, the user may want to transmit at five times this rate for short periods of time. If the bit rate on the ATM link is 155 Mbps, what is the sustained cell rate (SCR)? What is the peak cell rate (PCR)?

7.6 Assume that a VBR ATM protocol maps 45-byte segments of information plus 3 overhead bytes into the 48-byte ATM cell payload. Suppose that this ATM protocol receives data at 4 Mbps.
  (a) How many cells are created per second by the ATM layer?
  (b) What is the efficiency (ratio of bits received by the CBR protocol to bits sent out over the link) of ATM when using this protocol category?

7.7 Consider the case where three ONTs each are assigned 5 Mbps of fixed bandwidth over the same upstream FTTP link that can run at 622 Mbps. Suppose the ONTs send the following traffic: ONT1 sends two voice signals at 64 kbps each; every 10 seconds ONT2 sends 1-second bursts of information at a rate of 10 Mbps; ONT3 sends only idle cells at a rate of 10 per second.

(a) Describe the channel utilization for each ONT.

(b) How would the channel utilization change if a dynamic bandwidth allocation method were used?

7.8 The CRC used in the ATM header can correct all single errors and can detect all double errors occurring in the header. If bit errors occur in the header (of length $L = 40$ bits) at random with an error rate p, the probability $P(x)$ that x errors occur is given by

$$P(x) = \frac{L!}{x!(L-x)!}(1-p)^{L-x}p^x$$

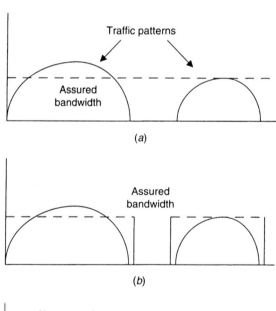

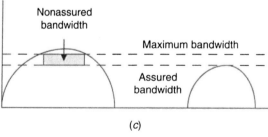

Figure 7.12. Bandwidth assignments for a general traffic pattern in terms of traffic container types: (a) T-CONT 1; (b) T-CONT 2; (c) T-CONT 3 (for Problem 7.13).

Find the probability that the header has no errors, one error, and two errors for error rates of $p = 10^{-6}$ and 10^{-9}.

7.9 (a) Find the binary equivalent of the polynomial $x^8 + x^7 + x^3 + x + 1$.

(b) Find the polynomial equivalent of 10011011110110101.

7.10 Consider the 10-bit data unit 1010011110 and the divisor 1011. Use both binary and algebraic division to find the CRC.

7.11 Consider the generator polynomial $x^3 + x + 1$.

(a) Show that the CRC for the data unit 1001 is given by 110.

(b) If the resulting code word has an error in the first bit when it arrives at the destination, what is the CRC calculated by the receiver?

7.12 A QAM constellation diagram consists of 64 phase and amplitude variations. If the bit rate is 48 Mbps, what is the symbol rate?

7.13 Consider the traffic pattern as a function of time shown in Figure 7.12. Describe the behavior of the three bandwidth assignments indicated by lines 1, 2, and 3 in terms of traffic container types T-CONT 1, T-CONT 2, and T-CONT 3.

7.14 For a data rate of 155 Mbps, to what transmission distance does a guard time of 4 bits correspond?

FURTHER READING

1. C. Bouchat, C. Dessauvages, F. Fredericx, C. Hardalov, R. Schoop, and P. Vetter, "WDM-upgraded PONs for FTTH and FTTBusiness," *Exp(Telecom Italia Lab)*, vol. 2, no. 2, pp. 6–17, July 2002; http://exp.telecomitalialab.com.
2. B. A. Forouzan, *Introduction to Data Communications and Networking*, 2nd ed., McGraw-Hill, Burr Ridge, IL, 2001.
3. G. Keiser, *Local Area Networks*, 2nd ed., McGraw-Hill, Burr Ridge, IL, 2002.
4. K. Kitayama, H. Mukai, K. Murakami, and T. Yokotani, "A study on usage of T-CONT in DBA for FTTx," *Exp(Telecom Italia Lab)*, vol. 2, no. 2, pp. 64–71, July 2002; http://exp.telecomitalialab.com.
5. A. Leon-Garcia and I. Widjaja, *Communication Networks*, McGraw-Hill, Burr Ridge, IL, 2000.
6. S. Ovadia, *Broadband Cable TV Access Networks, From Technologies to Applications*, Prentice Hall, Upper Saddle River, NJ, 2001.
7. J. Proakis, *Digital Communications*, 4th ed., McGraw-Hill, New York, 2001.
8. W. I. Way, *Broadband Hybrid Fiber/Coax Access System Technologies*, Academic Press, New York, 1998.

8

ETHERNET IN THE FIRST MILE

The extensive use of BPON technology to deploy the fiber-to-the-premises (FTTP) networks is driven by the fact that exchange carriers have an extensive embedded ATM switching infrastructure. Although ATM is a proven technology with scalable and flexible traffic-management capabilities and robust quality-of-service features, it does have several limitations. For example, ATM segments variable-sized IP packets into fixed-size cells with payloads of 48 bytes or less. The protocol then adds a 5-byte header, which results in a 10 percent cell-tax overhead (5 out of 53 bytes). In addition, a single corrupted or dropped ATM cell requires retransmission of the entire IP packet, even though other ATM cells belonging to the same IP packet may be received correctly. Such factors result in a large bandwidth overhead and a high demand on processing resources.

The widespread use of Ethernet in both local area and metro networks makes it an attractive alternative transport technology for access networks. Since this method encapsulates and transports data in Ethernet frames, it is easy to carry IP packets over an Ethernet link. This scheme thus simplifies the interoperability of metro and WAN assets with installed Ethernet LANs compared to the use of BPON technology.

In this chapter we first describe the general *Ethernet-in-the-first-mile* (EFM) concept and various ways to implement it. Next, the *Ethernet passive optical network* (EPON) operational concepts are presented in Section 8.2, including bidirectional traffic flows and packet encapsulation. Since an EPON uses a shared transmission medium, an essential network function is to have orderly and efficient traffic control. This is discussed in Section 8.3. The principles of point-to-point Ethernet are described in Section 8.4. Finally, in Section 8.5 we outline the constituents of the IEEE 802.3ah EFM standard and the ITU-T Recommendation G.985 for EPONs. Note that for simplicity of notation, in this chapter we use the acronym ONT to refer to both an ONT and an ONU.

FTTX Concepts and Applications, by Gerd Keiser
Copyright © 2006 John Wiley & Sons, Inc.

8.1 EFM OPTIONS

Ethernet has become the most widely used protocol for local area networks and has been extended into both metro and wide area networks. There is an installed base of over 500 million Ethernet ports, which has resulted in low-cost and highly reliable components that are widely available worldwide. Using Ethernet as a first-mile transport technology allows engineers to build access networks with IP and Ethernet, since data are carried in standard Ethernet frames. This avoids the cost and complexity of protocol conversion, as is done with BPON implementations. In addition, the deployment of Ethernet in the first mile enables network managers to take advantage of existing and familiar management and analysis tools that have been designed for monitoring and control of an Ethernet network.

Three different EFM physical transport schemes are possible. As illustrated in Figure 8.1, one uses an EPON methodology and the other two employ point-to-point (P2P) links over either copper wires or optical fibers that connect users and the central office directly. The architecture and operation of an EPON are described in Sections 8.2 and 8.3. In Section 8.4 we discuss the characteristics of different point-to-point links. In such a scenario there can be either many lines running to individual premises from the central office or a single fiber or copper link can connect the central office to a local distribution switch. Table 8.1 highlights the physical layer characteristics of these three options. They are spelled out in more detail in the IEEE

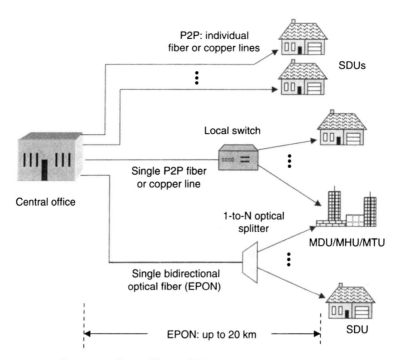

Figure 8.1. Three different EFM physical transport schemes.

TABLE 8.1 Main Physical Layer Characteristics of Three EFM Options

EFM Scheme	Physical Layer Options
EPON	10-km distance; 1 Gbps; 1 × 32 splitter; one bidirectional single-mode fiber
	20-km distance; 1 Gbps; 1 × 16 or 1 × 32 splitter; one bidirectional single-mode fiber
P2P over fiber	1000BASE-LX: extended temperature range optics for 1-Gbps links
	1000BASE-X: 10 km over one bidirectional single-mode fiber for 1-Gbps links
	100BASE-X: 10 km over one bidirectional single-mode fiber for 100-Mbps links
P2P over copper	750-m distance; 10 Mbps full-duplex transmission over single-pair nonloaded voice-grade copper wires

802.3ah EFM standard, which was approved in June 2004 (see Section 8.5). Two industry advocacy and standards supporting groups are the EPON Forum (http://www.ieeecommunities.org/epon) and the Ethernet in the First Mile Alliance (http://www.efmalliance.org).

8.2 EPON ARCHITECTURE

Figure 8.2 shows the basic EPON architecture and operational concept. This architecture follows the standard PON layout described in Chapter 6, which has one main feeder line going to an optical splitter. Up to 32 distribution branches leave

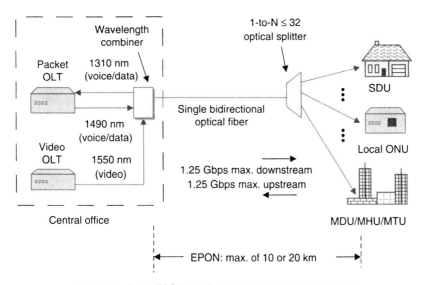

Figure 8.2. Basic EPON architecture and operational concept.

the splitter and interface to ONTs. As noted in Table 8.1, the IEEE 802.3ah EFM standard specifies operational conditions for a maximum transmission distance of either 10 or 20 km between the OLT and an ONT. The achievable transmission distance depends on the optical splitter size (16 or 32 subscriber ports) and on whether short-reach (class B) or long-reach (class C) optics are selected.

The EPON implementation uses commercially available Ethernet *media access control* (MAC) and *physical layer* (PHY) chip sets. This results in a significant economic benefit because of the wide availability and proven reliability of these components. Analogous to the BPON scheme, an EPON uses a 1490-nm wavelength for downstream transmission of voice and data to the ONTs and a 1310-nm wavelength for the upstream return path from an ONT to the OLT. Thus, the 1550-nm window is available for other services, such as multichannel video transmission from the OLT to the users. Since it is based on the standard gigabit Ethernet protocol, an EPON has a nominal bit rate of 1250 Mbps, which is sent using 8B10B encoding (see Appendix D for more details on mBnB encoding methods). In this 8B10B encoding scheme, two extra redundant bits are incorporated into every 8-bit block of data to provide adequate timing for signal recovery and to have error-monitoring features. In addition to the 256 8-bit data characters in an 8B10B code (since $2^8 = 256$), there are 13 special 10-bit control characters. These are used, for example, to indicate whether the information bits are for idle data, test messages, or frame delineation.

8.2.1 OLT and ONT/ONU Functions

Similar to its function in other PON architectures, in an EPON an OLT acts as the network controller. All communications take place between the OLT and the ONTs, so there is no direct interaction between ONTs on the same EPON. Some key functions of an OLT include the following:

1. Determines continuously if any ONTs have joined or left the network. This is known as a *discovery process* (see Section 8.3.1).
2. Controls the registration of newly joined ONTs.
3. Assigns varying amounts of upstream transmission bandwidth to each ONT (see Section 8.3.2).
4. Performs a *ranging process* to calculate the transmission time delay between an OLT and each ONT (see Section 8.3.3).
5. Generates time-stamped messages for global time reference purposes.

8.2.2 EPON Traffic Flows

EPON data are carried in standard Ethernet frames. Figure 8.3 shows the format of a standard Ethernet MAC frame, which can vary in length from 72 to 1526 bytes. The functions of the eight frame fields are as follows:

- *Preamble.* The frame begins with a 7-byte preamble that repeats the 8-bit pattern 10101010 seven times. This square-wave pattern alerts the receiving system

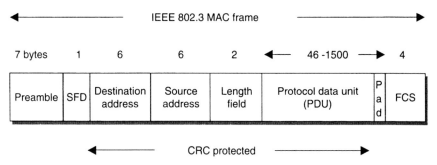

Figure 8.3. Format of a standard Ethernet frame.

that a frame is coming in and enables the receiver to synchronize its timing to the beginning of the frame.

- *Start frame delimiter* (SFD). The 1-byte SFD field consists of the sequence 10101011, where the two consecutive ones indicate the start of a frame. This enables the receiver to find the first bit of the frame.
- *Destination address* (DA). This 6-byte field contains the physical address of the next destination for the packet. The first DA bit distinguishes between addresses for a single user and addresses used to multicast a frame to a group of stations. The second bit tells whether this is a local or global address.
- *Source address* (SA). The 6-byte SA field contains the physical address of the last device that forwarded the packet. This could be the original sending station or the most recent router that received and forwarded the packet.
- *Length and type of PDU.* This 2-byte field indicates the number of bytes in the *protocol data unit* (PDU), which is the data field of the frame. Since the longest allowable Ethernet frame is 1526 bytes, the PDU can be up to 1520 bytes long.
- *Protocol data unit.* The PDU field contains logical link control (LLC) data and a variable-length information field. The function of the LLC is to provide address and control mechanisms to enable data exchange between end users.
- *Pad.* The pad field contains bytes that are added to ensure that the frame size is always at least 64 bytes long, which is the length required for proper collision-detection operation.
- *Frame check sequence* (FCS). This field contains the error-detection information for the frame. It is based on a 32-bit *cyclic redundancy check* (CRC) process that determines if there are errors in the frame. The CRC uses the bits in the DA, SA, length, and PDU fields to perform a binary division calculation at the sending station. The result of this calculation is added to the frame and is rechecked at the destination. If the results differ, an error has occurred in the frame during transmission and the frame is discarded.

In an EPON the preamble is not needed because of the full-duplex operational nature of the network. This factor is advantageous for embedding the ONT address

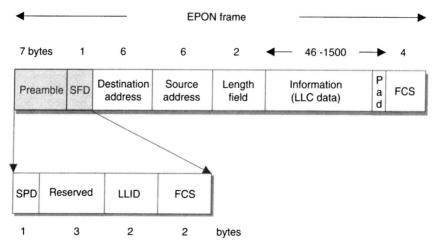

Figure 8.4. EPON frame format.

in the Ethernet frame. Thus, instead of increasing the Ethernet header size to accommodate an identification tag for indicating which ONT should accept the frame, this information replaces some of the bits in the standard Ethernet frame preamble. Consequently, the EPON frame has the same size and format as those of a standard Ethernet frame except that the preamble and SFD fields are replaced by the following EPON-related fields, as Figure 8.4 shows:

- A 1-byte *start-of-packet delimiter* (SPD) which contains clocking information. This synchronization marker is sent every 2 ms to synchronize the ONTs with the OLT.
- Three bytes reserved for future use.
- A 2-byte tag called a *logical link identifier* (LLID).
- A 2-byte *frame check sequence* (FCS) that contains error detection information for the EPON frame.

In the downstream direction, the OLT broadcasts Ethernet frames to the ONTs. This is compatible with the inherent broadcasting nature of the Ethernet protocol. An ONT receives and filters frames based on the LLID tag added to the frame by the OLT. For example, an ONT will reject frames intended for other ONTs and also discards broadcast frames that it has generated. When it sends data back to the OLT, an ONT places its own assigned LLID in the frame. Note that this LLID exists only within the EPON itself, since the ONT strips the LLID prior to sending the packet to the subscriber equipment.

A special procedure needs to be followed for traffic going in the upstream direction to avoid collisions between packets that might be transmitted simultaneously from different ONTs. That procedure is handled by the specially developed *multipoint control protocol* (MPCP), which arbitrates the channel access between the central office and the EPON subscribers. As described in Section 8.3 in more detail, one MPCP function is to assign the upstream bandwidth dynamically to various service subscribers. The MPCP procedure described in 802.3ah does not specify any particular dynamic bandwidth allocation algorithm, since the intent of the standard is to facilitate implementation of a DBA process.

8.2.3 Power Levels Received

As described in Chapter 4, in a PON the bursty nature of packet arrivals at the OLT results in large optical power level and phase variations in successive time slots. This level difference occurs because successive packets may come from different ONTs that are located at significantly varying distances from the OLT. Since conventional optical receivers are designed to accept optical signals with a constant amplitude and phase, they are not suitable for the burst-mode operation characteristic of PONs. It is therefore necessary to use *burst-mode receivers* which can adapt to the variation in optical power and phase alignment on a packet-by-packet basis. In Section 4.4 we describe this concept further.

Note that power-level variations are not a problem at the ONT, since the signals at any individual ONT arrive only from the OLT. However, similar to other PONs, the ONT lasers need to be turned off when not in use (see Section 4.5). Thus, in an EPON the ONTs use burst-mode transmitters but have conventional optical receivers.

8.3 MPCP FUNCTIONS

An EPON uses an MPCP to regulate both downstream and upstream traffic. The processes that the MPCP performs include auto-discovery, ONT registration, ranging, bandwidth polling, and bandwidth assignment. An EPON uses the following special types of 64-byte control messages for these functions:

1. GATE and REPORT messages are used to assign and request bandwidth.
2. The REGISTER_REQUEST, REGISTER, and REGISTER_ACK messages are used to implement the ONT auto-discovery, ONT registration, and ranging processes.

8.3.1 Discovery Process

An important necessity of an EPON is the ability to recognize the presence of an ONT and then to register it, negotiate operational parameters, assign an LLID, allocate upstream transmission bandwidth, and compensate for round-trip time delays. When an ONT is turned on or reset, it waits for a discovery GATE message containing a discovery LLID from the OLT. The OLT periodically broadcasts such a message to all the ONTs. The registration process then is carried out through a series of discovery GATE, REGISTER_REQUEST, REGISTER, and REGISTER_ACK messages. This message handshaking will register an ONT as belonging to the EPON with a uniquely assigned LLID and the necessary bandwidth time slots. Typically, an ONT is registered in less than 1 second.

8.3.2 Bandwidth Assignment

In an access network, subscribers tend to submit traffic to the networks in bursts. This is in contrast to a metro or wide area network, where the traffic flow tends to be relatively smooth, owing to the aggregation of many traffic sources. To handle this situation in a PON, the upstream bandwidth is divided into time-slot units. The OLT controls and assigns these units to the ONTs based on their transmission requirements. These units can be assigned dynamically as needed by an ONT, or they can be reserved in advance. By using a DBA, any reserved units or fraction of units of bandwidth that go unused can be reassigned dynamically by the OLT to other ONTs that may need it.

As shown in Figure 8.5, to help assign and request bandwidth in an EPON, the MPCP employs the 64-byte REPORT and GATE control messages. The ONT uses

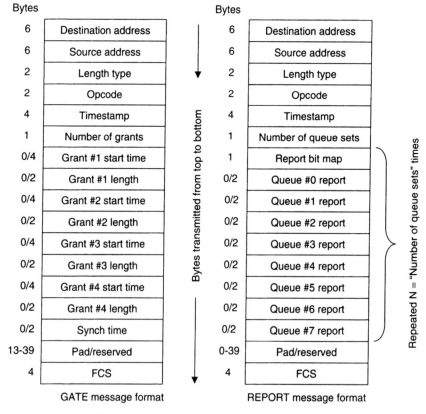

Figure 8.5. REPORT and GATE control messages.

the REPORT message to indicate its bandwidth requirements to the OLT. Typically, this is done in the form of queue occupancies, since each ONT has a set of buffers which hold queued Ethernet frames that are ready for upstream transmission to the OLT. A REPORT message from one ONT can indicate the status of up to eight queues, and each queue can have multiple thresholds. An ONT that reports the status of eight queues can have up to two thresholds per queue. If the ONT has only one queue, it is allowed to report up to 13 thresholds. When it receives a REPORT message, the OLT passes the message to the DBA algorithm. This algorithm then calculates the upstream transmission schedule of all the ONTs. Once the DBA system completes the upstream scheduling, the OLT transmits GATE messages to issue transmission grants to the ONTs. An individual GATE message can support up to four transmission grants.

8.3.3 Transmission Timing

For an ONT to transmit in a specific time slot, it needs to be synchronized to the OLT and to the other ONTs. This timing is done through an exchange of time-stamped synchronization control messages between the OLT and the ONT. Therefore, a transmission grant to an ONT also specifies the transmission start time and transmission length for that particular ONT. Each ONT updates its local clock using the timestamp contained in each transmission grant received. By this procedure, each ONT acquires and maintains global synchronization with the rest of the EPON.

Figure 8.6 illustrates the procedure for establishing global timing. First the OLT sends a GATE message at the absolute time T1, which is based on a global clock at the OLT. An ONT receives the GATE message at the local time T_{ONT2} and resets its local clock to read absolute time T1. The ONT then sends a REPORT message back

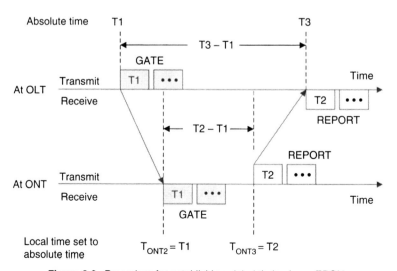

Figure 8.6. Procedure for establishing global timing in an EPON.

to the OLT at time T_{ONT3}, which now is absolute time T2. Therefore, the time stamp on this REPORT message reads T2. The OLT receives the REPORT message at absolute time T3. The round-trip time (RTT) then is given by

$$RTT = (T3 - T1) - (T2 - T1) = T3 - T2$$

This value is then used by the OLT to calculate the delay compensation time for the ONT.

8.4 POINT-TO-POINT ETHERNET

As an alternative to using an EPON in the access network, ITU-T Recommendation G.985 gives the specifications for a 100-Mbps *point-to-point* (P2P) Ethernet-based optical access system. In addition, IEEE 802.3ah describes networks that can use either optical fibers or voice-grade copper wires.

8.4.1 P2P Ethernet Over Fiber

Figure 8.7 shows one alternative for a point-to-point Ethernet optical access network. In this case there are dedicated fibers running between a central office and individual subscribers. Such a scenario requires a large number of optical fiber lines, with each line having its own optical transceivers. For example, suppose that the network serves 16 subscribers. If the optical link running to an individual subscriber is bidirectional, this scheme needs 16 fibers. In case the links are unidirectional, a total of 32 optical fibers are needed. Since each subscriber link needs transmitters and receivers at each end, the system needs a total of 32 optical transceivers. Therefore, this type of network is useful only if each subscriber requires close to the full capacity offered by a high-capacity Ethernet line.

The other option is to run either one bidirectional or two unidirectional fibers from the central office to an Ethernet switch located in the neighborhood of the subscribers, as Figure 8.8 shows. Individual fiber lines can then be run from the switch to each subscriber site. This network layout greatly reduces the number of fibers

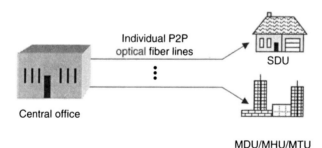

Figure 8.7. One alternative for a point-to-point Ethernet optical access network.

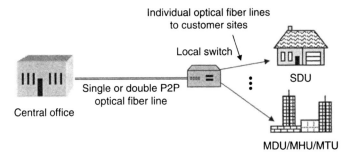

Figure 8.8. P2P EFM option with one bidirectional or two unidirectional fibers and a local Ethernet switch.

interfacing to the central office, but now the use of an Ethernet switch requires electrical power in the outside cable plant. Also, in addition to the 32 optical transceivers used for links between the subscribers and the local Ethernet switch, either two or four more optical transceivers are needed for the link running between the central office and the switch.

In contrast to these P2P network configurations, an EPON serving 16 ONTs requires one fiber from the central office to a local splitter, 17 optical transceivers, and no electrical power in the outside cable plant.

8.4.2 P2P Ethernet Over Copper

In addition to P2P optical links and an EPON specification, the IEEE 802.3ah standard also addresses the use of a single pair of nonloaded, voice-grade copper lines for access network links. The term *nonloaded* means that there are no special signal conditioning devices in the copper line. The standard specifies full-duplex transmission distances up to 750 m with maximum data rates of 10 Mbps. The reason behind this option is that a large percentage of the installed outside cable plant media in access networks consists of voice-grade, single-pair copper wires, and much of the installed media inside buildings is category 1, 2, or 3 twisted-pair copper wires.

8.5 MAIN EPON AND P2P EFM STANDARDS

The IEEE Standard 802.3ah for Ethernet-based access network architectures was approved in July 2004. The standard addresses the following three areas: point-to-point links over existing copper wires at 10-Mbps speeds over distances up to 750 m, point-to-point optical fiber at speeds up to 1 Gbps over distances up to 10 km, and point-to-multipoint optical fiber links at the 1-Gbps maximum speeds over maximum distances of 20 km. This last area encompasses Ethernet passive optical network implementations with either 1-to-16 or 1-to-32 optical power splitting ratios. In addition, procedures for operation, administration, and maintenance (OAM) are included to support network operation and troubleshooting. To assist with the OAM functions, options for unidirectional transmission of frames are provided for 100BASE-X, 1000BASE-X, 10GBASE-R, 10GBASE-W, and 10GBASE-X applications. In these designations, the numbers 100 and 1000 refer to the transmission speed in Mbps, and 10G means 10 Gbps rates. The letters X, R, and W refer to the basic physical layer families defined in the standard. The X family designates the use of 8B10B encoding, and the letter R designates the use of 64B66B encoding of the data. The W family specifies the encapsulation of 64B66B encoded Ethernet media-access-control data within a SONET or SDH frame.

The ITU-T Recommendation G.985 for 100-Mbps point-to-point Ethernet optical access systems was approved in March 2003. Included in this document are specifications for the optical distribution network (ODN), the physical layer, and OAM processes. The specifications are based on using a one-fiber wavelength-division-multiplexed bidirectional transmission system.

8.6 SUMMARY

The widespread use of Ethernet in both local area and metro networks makes it an attractive transport technology for access networks. Since this method encapsulates and transports data in Ethernet frames, it is easy to carry IP packets over an Ethernet link. This scheme thus simplifies the interoperability of metro and WAN assets with installed Ethernet LANs compared to the use of BPON technology. The general concept for implementing Ethernet in an access network is referred to as Ethernet-in-the-first-mile (EFM).

Three different EFM physical transport schemes are possible. One uses an EPON methodology and the other two employ point-to-point links over either copper wires or optical fibers that connect users and the central office directly. Table 8.1 lists the main characteristics of these three options. They are spelled out in more detail in the IEEE 802.3ah EFM standard. Two industry advocacy and standards supporting groups are the EPON Forum (http://www.ieeecommunities.org/epon) and the Ethernet in the First Mile Alliance (http://www.efmalliance.org).

Analogous to the BPON scheme, an EPON uses a 1490-nm wavelength for downstream transmission of voice and data to the ONTs and a 1310-nm wavelength for the upstream return path from an ONT to the OLT. The 1550-nm window is available for other services, such as multichannel video transmission from the OLT to the users.

PROBLEMS

8.1 A 3B4B code converts blocks of 3 bits to blocks of 4 bits according to the rules given in Table 8.2. When there are two or more consecutive blocks of three zeros, the coded binary blocks 0010 and 1101 are used alternately.

TABLE 8.2 3B4B Code Conversion Rules

| Original | 3B4B Code Word | |
Code Word	Mode 1	Mode 2
000	0010	1101
001	0011	
010	0101	
011	0110	
100	1001	
101	1010	
110	1100	
111	1011	0100

Similarly, the coded blocks 1011 and 0100 are used alternately for consecutive blocks of three ones.

(a) Using these translation rules, find the coded bit stream for the data input

010001111111101000000001111110.

(b) What is the maximum number of consecutive identical bits in the coded pattern?

8.2 A 4B5B code has $2^4 = 16$ four-bit data characters. The code maps these into the 5-bit sequences listed in Table 8.3. Using this information, encode the bit stream

010111010010111010100111

TABLE 8.3 Data Sequences Used in 4B5B Code Conversion

Data Sequence	Encoded Sequence	Data Sequence	Encoded Sequence
0000	11110	1000	10010
0001	01001	1001	10011
0010	10100	1010	10110
0011	10101	1011	10111
0100	01010	1100	11010
0101	01011	1101	11011
0110	01110	1110	11100
0111	01111	1111	11101

8.3 Consider an Ethernet link in which a station is transmitting frames with a 1500-byte information field continuously. Suppose that a 1-ms data-corrupting electric pulse occurs on the line. How many frames would be corrupted for Ethernet rates of 10 and 100 Mbps?

8.4 What are the round-trip times for messages sent over a fiber optic link from an OLT to ONTs that are located 3 and 20 km away?

8.5 Suppose that a telecommunications company needs to service two clusters of eight subscribers. As Figure 8.9 shows, each cluster is 10 km from the central office and the distance between the two clusters is 10 km. Describe some advantages and limitations to using

(a) an EPON architecture to connect all 16 subscribers and

(b) point-to-point Ethernet links running from the central office to each subscriber.

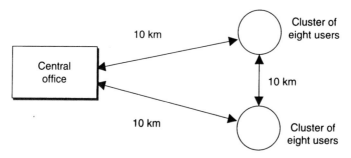

Figure 8.9. Network architecture for Problems 8.5 and 8.6.

8.6 For the network shown in Figure 8.9, describe how the following services might be implemented between users in the two clusters:

(a) requests for and exchanges of large files;

(b) a videoconferencing session.

8.7 Write a one-page trade-off between BPON and EPON technologies.

8.8 Investigate the operational details of the multipoint control protocol (MPCP). Consider factors such as bandwidth request and assignment, negotiation of parameters, managing and timing upstream transmissions to avoid collisions, ranging, and auto-discovery and registration of ONTs.

8.9 Using vendor resources (e.g., such as can be found on the Internet), describe the operational characteristics of typical OLT and ONT equipment that can be used for EPON links. Consider parameters for an OLT such as automatic ONT provisioning, LLID assignments, dynamic bandwidth allocation, ability to assign operational parameter values based on service level agreements, and element management capabilities.

8.10 Using vendor resources (e.g., such as can be found on the Internet), describe the operational characteristics of Ethernet transceivers that can be used for point-to-point links in an access network. Consider parameters such as size, transmission speed, transmission distances, hot-swappable characteristics, and optical connector interfaces.

FURTHER READING

1. G. Held, *Ethernet Networks: Design, Implementation, Operation, Management*, 4th ed., Wiley, Hoboken, NJ, 2003.
2. IEEE 802.3ah, "Ethernet in the First Mile Standard," July 2004.
3. ITU-T Recommendation G.985, "100 Mbit/s Point-to-Point Ethernet Based Optical Access System," Mar. 2003.
4. G. Keiser, *Local Area Networks*, 2nd ed., McGraw-Hill, Burr Ridge, IL, 2002.
5. G. Kramer, *Ethernet Passive Optical Networks*, McGraw-Hill, New York, 2005.
6. G. Kramer and G. Pesavento, "Ethernet passive optical network (EPON): building a next-generation optical access network," *IEEE Commun. Mag.*, vol. 40, pp. 66–73, Feb. 2002.
7. G. Kramer, B. Mukherjee, and A. Maislos, "Ethernet Passive Optical Networks," in S. Dixit, ed., *IP over WDM: Building the Next Generation Optical Internet*, Wiley, Hoboken, NJ, 2003.
8. M. P. McGarry, M. Maier, and M. Reisslein, "Ethernet PONs: a survey of dynamic bandwidth allocation (DBA) algorithms," *IEEE Commun. Mag.*, vol. 42, pp. S8–S15, Aug. 2004.
9. J. Xie, S. Jiang, and Y. Jiang, "A dynamic bandwidth allocation scheme for differentiated services in EPONs," *IEEE Commun. Mag.*, vol. 42, pp. S32–S39, Aug. 2004.

9

GPON CHARACTERISTICS

The growing demand for higher speeds in the access network, and the widespread use of both ATM and Ethernet, spawned the idea of developing a PON with capabilities beyond those of the BPON and EPON architectures. A major aim of this idea was to develop a versatile PON with a frame format that could transmit variable-length packets efficiently at gigabit per second rates. The FSAN group started such an effort in April 2001. The result was the ITU-T Recommendation series G.984.1 through G.984.4 for a gigabit PON (GPON).

In this chapter, in Section 9.1 we address the GPON framework as given in the ITU-T Rec. G.984.1. This framework, which is known as the *GPON service requirements* (GSR), summarizes the operational characteristics that service providers expect of the network. Next, in Section 9.2 we describe the *GPON physical medium dependent* (GPM) specifications given in ITU-T Rec. G.984.2. This includes operational parameters of the optical transmitters and receivers, clock recovery, and error-correction mechanisms. Section 9.3 covers the *GPON transmission convergence* (GTC) specifications as given in ITU-T Rec. G.984.3. The GTC is responsible for correct implementation of the data flow process in the physical layer and addresses issues such as the frame structure, the control sequence between the OLT and the ONTs, and the packet encryption function. ITU-T Rec. G.984.4 defines the *ONT management and control interface* for a GPON. An overview of this specification is given in Section 9.4.

9.1 GPON ARCHITECTURE

As shown in Figure 9.1, the network layout for a GPON follows that of the standard PON concept described in Chapter 6. It also retains many of the same functionalities of the BPON and EPON schemes, such as dynamic bandwidth assignment (DBA) and the use of operations, administration, and maintenance (OAM) messages.

FTTX Concepts and Applications, by Gerd Keiser
Copyright © 2006 John Wiley & Sons, Inc.

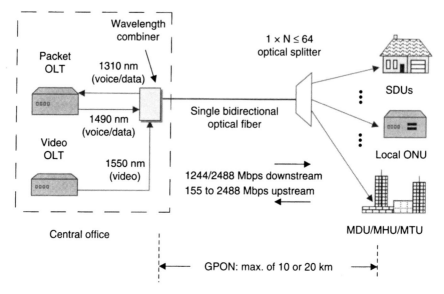

Figure 9.1. General GPON architecture and characteristics.

However, in contrast to BPON and EPON architectures, which were developed from an equipment vendor point of view, the GPON operational scheme is a more customer-driven design. This is reflected in the GPON service requirements specification described in G.984.1. This document takes into account the collective requirements of the leading communication service providers in the world.

The G.984.1 recommendation was approved by the ITU-T in March 2003. It describes general characteristics such as the GPON architecture, what types of services to accommodate, the desired bit rates, signal transfer delays, optical power-splitting ratios, and information security. Note that to be backwardly compatible with existing BPON systems, G.984.1 maintains some requirements of the BPON G.983.1 Recommendation.

9.1.1 GSR Specification

Table 9.1 highlights the major GSR specifications. First a GPON must be a *full-service network*, which means that it should be able to carry all service types. These include 10- and 100-Mbps Ethernet, legacy analog telephone, digital T1/E1 traffic (i.e., 1.544 and 2.048 Mbps), 155-Mbps asynchronous transfer mode (ATM) packets, and higher-speed leased-line traffic. The nominal line rates are specified as 1.25 Gbps (1244.160 Mbps) and 2.5 Gbps (2488.320 Mbps) in the downstream direction and 155 Mbps, 622 Mbps, 1.25 Gbps, and 2.5 Gbps in the upstream direction. The data rates can be either symmetrical (the same rate in both directions) or asymmetrical, with higher rates being sent downstream from the OLT to the ONTs. A service provider can offer a lower upstream rate to those GPONs in which the

TABLE 9.1 Summary of GPON Service Requirements

Parameter	GSR Specification
Service	Full service: for example, 10/100 BASE-T Ethernet, analog telephony, SONET/SDH TDM, ATM
Access data rate	Downstream: 1.244 and 2.488 Gbps; upstream: 155 Mbps, 622 Mbps, 1.244 Gbps, 2.488 Gbps
Distance	10 or 20 km maximum
Splitting number	Maximum of 64
Wavelengths	Downstream voice/data: 1480 to 1500 nm; upstream voice/data: 1260 to 1360 nm; downstream video distribution: 1550 to 1560 nm
Protection switching	Fully redundant 1 + 1 protection; partially redundant 1 : N protection
Security	Information security at the protocol level for downstream traffic: for example, the use of the *Advanced Encryption Standard* (AES)

downstream traffic is much larger than in the upstream direction, as is the case when subscribers use the IP data service mainly for applications such as lower-rate upstream Internet surfing or e-mail and higher-rate downstream downloads of large files.

The wavelengths are specified to be in the range 1480 to 1500 nm for downstream voice and data traffic and 1260 to 1360 nm for its corresponding upstream traffic. Thus, the median values are the standard 1490- and 1310-nm wavelengths as used in BPON and EPON systems. In addition, the wavelength range 1550 to 1560 nm can be used for downstream video distribution. Depending on the capabilities of the optical transmitters and receivers, the GPON recommendation specifies maximum transmission distances of 10 or 20 km. For a GPON the maximum number of splitting paths is 64.

9.1.2 GPON Protection Switching

The ITU-T Recommendation G.984.1 describes the use of a protection switching mechanism that is compatible with BPON operation. This allows several different types of PON configurations, including redundancy of links and equipment for network protection. Among these are a fully redundant 1 + 1 protection and a partially redundant 1 : N protection. Figure 9.2 shows that in 1 + 1 protection the traffic is transmitted simultaneously over two separate fiber lines from the source to the destination. Typically, these two paths do not overlap at any point, so that a cable cut would

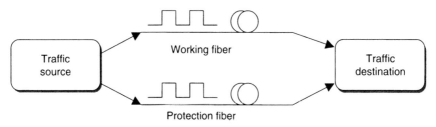

Figure 9.2. Fully redundant 1 + 1 protection of links.

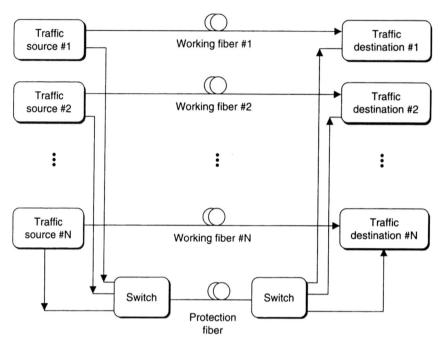

Figure 9.3. The 1 : N protection procedure.

affect only one fiber transmission path. In the 1 + 1 protection scheme, the receiving equipment selects one of the links as the *working fiber* for reception of information. In case a fiber in that link is cut or the transmission equipment on that link fails, the receiver switches over to the *protection fiber* and continues to receive information. This protection method provides rapid switchover during failures and does not require a protection signaling protocol between the source and destination. However, it requires duplicate fibers and redundant transmission equipment for each link.

The 1 : N protection procedure offers a more economical use of fibers and equipment. As shown in Figure 9.3, here one protection fiber is shared among N working fibers. This arrangement offers protection in the event that one of the working fibers fails. For most operational networks this level of protection is adequate, since failures of multiple fibers are rare (unless all the fibers are in the same cable). In contrast to the 1 + 1 protection method, in the 1 : N protection scheme traffic is transmitted only over the working fiber in normal operation. When there is a failure on a particular link, the source and destination both switch over to the protection fiber. This requires an automatic switching protocol between the endpoints to enable use of the protection link.

9.1.3 Information Security in a GPON

As is the case with other PON architectures, since the downstream data from the OLT is broadcast to all ONTs, every message transmitted can be seen by all the users attached to the GPON. Thus, the GPON standard describes the use of an information security mechanism to ensure that users are allowed to access only the data intended for them. In addition, such a security mechanism ensures that no malicious eavesdropping threat is probable. One example of a point-to-point encryption mechanism is the *Advanced Encryption Standard* (AES), which is used to protect the information payload of the data field in the GPON frame. *Encryption* is a technique whereby data are transformed into an unintelligible format at the sending end to protect them against unauthorized disclosure, modification, utilization, or destruction as they travel through the network.

The AES algorithm encrypts and decrypts 128-bit data blocks from their original format called *plaintext* to an unintelligible form called *ciphertext*. The cipher keys can have lengths of 128, 192, or 256 bits, which makes the encryption extremely difficult to compromise. A key can be changed periodically (e.g., once per hour) without disturbing the information flow. Decrypting the ciphertext converts the data back into its original form.

9.2 GPON RECOMMENDATION G.984.2

The ITU-T Recommendation G.984.2 describes the requirements and specifications for the GPON physical-media-dependent layer, which is equivalent to the physical layer of the OSI reference model. The specifications include electrical-to-optical and optical-to-electrical conversions, clock recovery, and error-correction mechanisms. Table 9.2 summarizes the key GPM specifications.

9.2.1 Optical Performances

A main concern of G.984.2 is to specify the optical transmitter and receiver performances at 10- and 20-km transmission distances. For power-budget calculations, GPON uses the same classes of optics as specified for BPON systems. These optical power budgets are:

- *Class A optics*: 5 to 20 dB
- *Class B optics*: 10 to 25 dB
- *Class C optics*: 15 to 30 dB

TABLE 9.2 Summary of Key GPON Physical-Medium-Dependent Specifications

Parameter	GPM Specification
Access data rate	Downstream: 1.244 and 2.488 Gbps; upstream: 155 Mbps, 622 Mbps, 1.244 Gbps, 2.488 Gbps
Optics classes	Classes A, B, and C; same requirements as for BPON systems
Burst overhead	Burst overhead is specified at each transmission speed
Power leveling	The ONT optical output can operate in three power-level modes to relieve photodetector tolerance at the OLT
Data reliability	Forward error correction (FEC) may be used as an option

Included in these attenuation ranges are losses resulting from the optical fiber, splices, connectors, and optical splitters. In addition, the link designer needs to take into account the possible occurrence of other link degradations, such as additional splices and fiber lengths resulting from cable repairs, the effects of environmental factors on cable performance, and unforeseen degradations in any passive components. In Section 6.6 we provide more details on the longest transmission distance achievable under ideal conditions with high-quality components for the maximum power budgets within the three different optical network classes.

9.2.2 Timing and Optical Power Control

An important ingredient of G.984.2 is the specification of the *burst-mode timing* parameters. Even though a GPON is a synchronous network in which the OLT informs the ONTs when they can transmit, this process has some timing uncertainty. Although this also is the case for the BPON and EPON schemes, the higher data rates (and therefore the smaller pulse widths in a packet burst) make this factor even more critical in a GPON.

Since the ranging procedure has a limited accuracy, a guard time is placed between consecutive bursts from the ONTs to avoid collisions of the independent packets. Figure 9.4 illustrates this arrangement. The guard time interval has to account for factors such as the laser turn-on delay of the next information burst, the turnoff delay of the laser and the residual tail current of the preceding burst due to component discharge times in the OLT receiver, and the granularity of time-delay equalization among the various ONTs in the GPON. The guard time nominally is 25.6 ns, which means that the number of bits allocated to that field becomes larger as the data rate increases. For example, a 25.6-ns guard time consumes 16 bits at 622 Mbps, 32 bits at 1244 Mbps, and 64 bits at 2.488 Gbps. In Section 9.3.2 we give further details on the structure of the physical layer overhead field that addresses these factors.

If an ONT is located close to the OLT, the avalanche photodetector in the OLT receiver might see a relatively high optical power level from that ONT, as illustrated in Figure 9.5. To relieve the OLT detector from experiencing possible overload conditions, G984.2 allows for *optical power control* in the ONT transmitter through the implementation of three power-level modes. In mode 1 the ONT transmitter operates at its normal optical output level. In modes 2 and 3 the ONT optical output is 3 and 6 dB

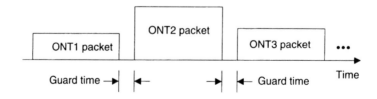

Figure 9.4. A guard time between consecutive bursts avoids packet collisions.

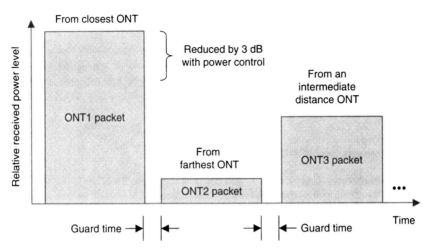

Figure 9.5. Successive packets arriving at an OLT from different ONTs may have large optical power-level variations.

lower than mode 1, respectively. Thus if the OLT photodetector receives too strong a power level, the OLT can command the sending ONT to reduce its optical output level.

9.2.3 Forward Error Correction

As an option aimed at keeping the costs of the optical transceivers low, yet maintaining a high level of data integrity, G984.2 specifies the use of *forward error correction* (FEC). The application of FEC in a GPON is desirable since the higher bit rates reduce the optical power budget in two ways. First, higher bit rates reduce receiver sensitivity since the wider bandwidth of the receiver introduces more noise. The other reason is that chromatic dispersion effects are greater at higher bit rates, which results in a larger power penalty along the optical path. Thus, using FEC is helpful, since it can add about 3 to 6 dB to the power budget.

FEC is a mathematical signal-processing technique that encodes data so that errors can be detected and corrected. In FEC techniques, redundant information is transmitted along with the original information. If some of the original data are lost or received in error, the redundant information is used to reconstruct the data. Typically, the amount of redundant information is small, so the FEC scheme does not use up much additional bandwidth and thus remains efficient.

The most popular error-correction codes are *cyclic codes*. These are designated by the notation (n, m), where n equals the number of original bits m plus the number of redundant bits. Although many types of cyclic codes have been considered over the years in electrical systems, the *Reed–Solomon* (RS) FEC codes are among the best suited for optical signals. They have a low overhead, high coding gain, and can correct bursts of errors. The RS (255, 239) code is used widely for telecom lines. This code changes 239 data bits into 255 transmitted bits, thereby adding about 7 percent overhead.

9.3 GPON TRANSMISSION CONVERGENCE LAYER

The ITU-T Recommendation G.984.3 describes the *transmission convergence* (TC) layer, which is equivalent to layer 2 (the data transmission layer) in the OSI reference model. It specifies the GPON frame format, the media access control protocol, the ranging scheme, operations and maintenance processes, and the information encryption method.

9.3.1 Downstream GPON Frame Format

To accommodate all types of services (e.g., ATM, TDM, and Ethernet) efficiently, a *GPON encapsulation method* (GEM) is used. This method is based on a slightly modified version of the ITU-T Recommendation G.7041 *Generic Framing Procedure*, which gives the specifications for sending IP packets over SONET or SDH networks.

Figure 9.6 shows the GPON frame format, which has a fixed 125-μs length. The frame consists of a *physical control block* (PCB) and a payload composed of a pure ATM segment and a GEM segment. The PCB section contains the physical layer overhead information used to control and manage the network.

In the downstream direction the PCBd (physical control block for frames going downstream) contains the following information:

- A 4-byte frame synchronization field (Psync).
- A 4-byte segment (Ident) that contains an 8-kHz counter, a downstream FEC status bit, an encryption key switchover bit, and 8 status bits reserved for further use.
- A 13-byte downstream physical layer OAM (PLOAMd) message, which handles functions such as OAM-related alarms or threshold-crossing alerts.

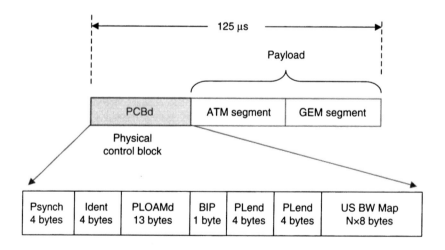

Figure 9.6. The three-segment GPON frame format has a 125-μs length.

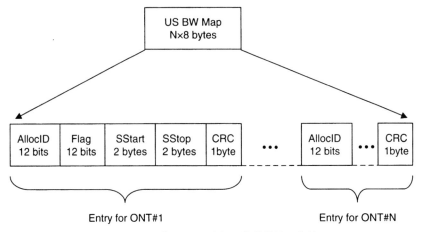

Figure 9.7. Structure of the US BW Map field.

- A 1-byte bit interleaved parity (BIP) field, used to estimate the bit error rate.
- A 4-byte downstream payload length indicator (PLend), which gives the length of the upstream bandwidth (US BW) map and the size of the ATM segment. The PLend field is sent twice for extra redundancy and error robustness.
- The $N \times 8$-byte US BW map allocates N transmission time slots to the ONTs.

The US BW map contains N entries associated with N time-slot allocation identifications for the ONTs. As Figure 9.7 shows, each entry in the US BW Map or access structure consists of:

- A 12-bit *allocation identifier* (AllocID) that is assigned to an ONT.
- Twelve flag bits that allow the upstream transmission of physical layer overhead blocks for a designated ONT(see Section 9.3.2).
- A 2-byte *start pointer* (SStart) that indicates when the upstream transmission window starts. This time is measured in bytes; the beginning of the upstream GTC frame is designated as time zero.
- A 2-byte *stop pointer* (SStop) that indicates when the upstream transmission window stops.
- A 1-byte CRC that provides 2-bit error detection and 1-bit error correction on the bandwidth allocation field.

Figure 9.8 gives an example of time-slot allocations for three ONTs. Here there are three entries in the US BW Map field. The AllocID of the ONTs are 1, 2, and 3 for ONT1, ONT2, and ONT3, respectively. The center part of Figure 9.8 shows start and stop time slots listed in the downstream US BW Map field during which the various

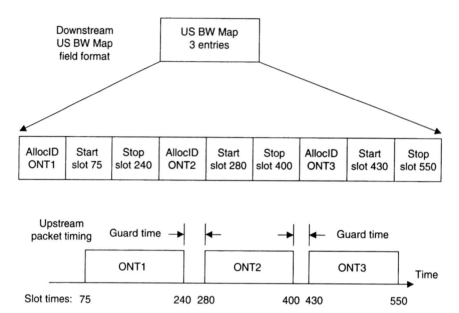

Figure 9.8. Downstream US BW Map field and the ensuing upstream response.

ONTs are allowed to transmit. The lower part of Figure 9.8 shows the general format of the ensuing upstream information stream from the three ONTs. An appropriate guard time is placed between packets from different ONTs.

9.3.2 Upstream GPON Frame Format

Upstream GPON traffic consists of successive transmissions from one or more ONTs. As Figure 9.8 illustrates, the particular sequence of frames is based on the transmission time-slot allocations developed by the OLT. To allow proper reception of the individual burst-mode frames, a certain amount of *burst overhead* is needed at the start of an ONT upstream burst. Figure 9.9 shows the format of an upstream frame, which consists of up to four types of PON overhead fields and a variable-length user data payload that contains a burst of transmission. The upstream header fields are the following:

- The *physical layer overhead* (PLOu) at the start of an ONT upstream burst contains the preamble, which ensures proper physical layer operation (e.g., bit and byte alignments) of the burst-mode upstream link.
- The upstream *physical layer operation, administration, and management* (PLOAMu) field is responsible for management functions such as ranging, activation of an ONT, and alarm notifications. The 13-byte PLOAMu contains the PLOAM message as defined in G.983.1 and is protected against bit errors by a *cyclic redundancy check* (CRC) that uses a standard polynomial error detection and correction code.

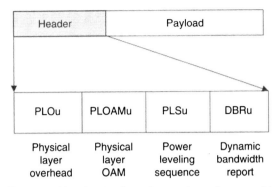

Figure 9.9. Format and header structure of an upstream frame sent from an ONT.

- The upstream *power leveling sequence* (PLSu) field contains information about the laser power levels at the ONTs as seen by the OLT. Given this information, the OLT uses the PLSu to adjust the ONT power levels to reduce the optical dynamic range received at the OLT.
- The *dynamic bandwidth report* (DBRu) field informs the OLT of the queue length of each AllocID at an ONT. This allows the OLT to enable proper operation of the dynamic bandwidth allocation process. The DBRu is protected against bit errors by a CRC.

Transmission of the PLOAMu, PLSu, and DBRu fields are optional depending on the downstream flags in the US BW map.

9.3.3 GEM Segment

The GPON encapsulation method works similar to ATM, but it uses variable-length frames instead of fixed-length cells as in ATM. Thus, GEM provides a generic means to send different services over a GPON. The encapsulated payload can be up to 1500 bytes long. If an ONT has a packet to send that is larger than 1500 bytes, the ONT must break the packet into smaller *fragments* that fit into the allowed payload length. The destination equipment is responsible for reassembling the fragments into the original packet format.

Figure 9.10 shows the GEM segment structure, which consists of four header fields and a payload that is L bytes long. The header fields are the following:

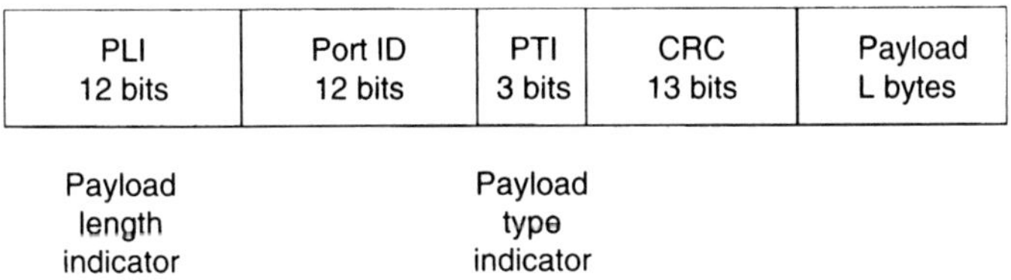

PLI 12 bits	Port ID 12 bits	PTI 3 bits	CRC 13 bits	Payload L bytes
Payload length indicator		Payload type indicator		

Figure 9.10. Four header fields and payload of the GEM segment.

- A 12-bit *payload length indicator* (PLI) that gives the length in bytes of the GEM-encapsulated payload.
- A 12-bit *port identification number* that tells which service flow this fragment belongs to.
- A 3-bit *payload type indicator* (PTI) which specifies if the fragment is the end of a user datagram, if the traffic flow is congested, or if the GEM payload contains OAM information.
- A 13-bit cyclic redundancy check for header error control that enables the correction of two erroneous bits and the detection of three bit errors in the header.

A key advantage of the GEM scheme is that it provides an efficient means to encapsulate and fragment user information packets. The reason for using encapsulation on a GPON is that it allows proper management of the multiple service flows from different ONTs that share a common optical fiber transmission link. The purpose of fragmentation is to send packets from a user efficiently regardless of their size and to recover the original packet format reliably from the physical layer transmission windows on the GPON.

9.4 ONT MANAGEMENT AND CONTROL

The ITU-T Recommendation G.984.4 defines the *ONT management and control interface* (OMCI) for a GPON. The OMCI for a GPON is basically identical to that for a BPON. The main area that is different is the management of the GPON encapsulation method. However, since the GEM is a connection-oriented protocol, it can be managed using the same entities and methodologies as are used for the ATM transport service.

9.5 SUMMARY

The concept of a gigabit passive optical network (GPON) addresses the demand for higher speeds in the access network beyond the capabilities offered by BPON and EPON architectures. In addition, it enables the transmission of both ATM cells and Ethernet packets in the same transmission frame structure. A major new feature in a GPON is a versatile frame format that can transmit variable-length packets efficiently at gigabit per second rates. The GPON network layout follows that of a standard PON concept. It also retains much of the same functionality characteristic of BPON and EPON schemes, such as dynamic bandwidth assignment (DBA) and the use of operations, administration, and maintenance (OAM) messages. However, in contrast to BPON and EPON architectures, which were developed from an equipment vendor point of view, the GPON operational scheme is a more customer-driven design.

A GPON is a full-service network, which means that it should be able to carry all services. These include 10- and 100-Mbps Ethernet, legacy analog telephone, digital T1/E1 traffic (i.e., 1.544 and 2.048 Mbps), 155-Mbps asynchronous transfer mode (ATM) packets, and higher-speed leased-line traffic. The nominal line rates are specified as 1.2 and 2.4 Gbps in the downstream direction and 155 Mbps, 622 Mbps, 1.2 Gbps, and 2.4 Gbps in the upstream direction. The data rates can be either symmetrical (the same rate in both directions) or asymmetrical, with higher rates being sent downstream from the OLT to the ONTs. A service provider can offer a lower upstream rate to those GPONs in which the downstream traffic is much larger than in the upstream direction, as is the case when subscribers use the IP data service mainly for applications such as lower-rate upstream Web surfing or e-mails and higher-rate downstream downloads of large files.

• The GPON Recommendation G.984.1 describes the use of a protection switching mechanism, which includes redundancy of links and equipment for network protection. Among these are a fully redundant $1 + 1$ protection scheme and a partially redundant (shared) $1 : N$ protection layout.

As is the case with other PON architectures, since the downstream data from the OLT are broadcast to all ONTs, every message transmitted can be seen by all the users attached to the GPON. Thus, the GPON standard describes the use of a security mechanism to ensure that users are allowed to view only the information intended for them. In addition, such a security mechanism ensures that no malicious eavesdropping threat is probable. One example of a point-to-point encryption mechanism is the Advanced Encryption Standard (AES), which is used to protect the information payload of the data field in the GPON frame.

An important ingredient of G.984.2 is the specification of the burst-mode timing parameters. Since the ranging procedure has a limited accuracy, a guard time is placed between consecutive bursts from the ONTs to avoid collisions of the independent packets. The guard time interval has to account for factors such as the laser turn-on delay of the next information burst, the turnoff delay of the laser and the residual tail current of the preceding burst due to component discharge times in the OLT receiver, and the granularity of time-delay equalization among the various ONTs in the GPON. In addition, to protect the OLT detector from possible overload conditions, three levels of optical power control may be used in the ONT transmitter.

To accommodate all types of services (e.g., ATM, TDM, and Ethernet) efficiently, a GPON encapsulation method (GEM) is used. This method is based on a slightly modified version of the ITU-T Recommendation G.7041 Generic Framing Procedure, which gives the specifications for sending IP packets over SONET or SDH networks.

PROBLEMS

9.1 Draw network diagrams of how a $1 + 1$ protective method and a $1 : 1$ arrangement could be implemented for a GPON. Describe how each configuration would function.

TABLE 9.3 Hexadecimal Representation of Bit Patterns for Problem 9.2

Bit Pattern	Character	Bit Pattern	Character
0000	0	1000	8
0001	1	1001	9
0010	2	1010	a
0011	3	1011	b
0100	4	1100	c
0101	5	1101	d
0110	6	1110	e
0111	7	1111	f

9.2 The basic unit for processing the Advanced Encryption Standard (AES) algorithm is the byte. For convenience, byte values are denoted using hexadecimal notation with each of two groups of 4 bits denoted by a single character, as shown in Table 9.3. For example, the element {01100011} can be represented by {63}. Using Table 9.3, find the binary equivalent of the following words:
(a) 32 43 f6 a8 88 5a 30 8d 31 31 98 a2 e0 37 07 34
(b) 2b 7e 15 16 28 ae d2 a6 ab f7 15 88 09 cf 4f 3c

9.3 Since the ranging procedure has a limited accuracy, a guard time is placed between consecutive bursts from the ONTs to avoid collisions of the independent packets, as Figure 9.4 illustrates. Verify that a 25.6-ns guard time consumes 16 bits at 622 Mbps, 32 bits at 1244 Mbps, and 64 bits at 2.488 Gbps.

9.4 List a sequence of steps that might be followed at a GPON OLT for implementing an optical power control process in an ONT (see Section 9.2.2). Assume that there are two optical power threshold levels in the OLT receiver.

9.5 (a) What is the overhead of the RS(255, 239) cyclic code?
(b) What impact does such a code have on GPON link performance?

9.6 (a) If the average GEM frame payload length for transporting IP (Internet protocol) data is 200 bytes, what is the GEM efficiency?
(b) Given that Ethernet encapsulates each frame with 21 bytes, what is the efficiency of Ethernet for transporting IP data if the payload is 200 bytes?

9.7 Using vendor resources (e.g., such as can be found on the Internet), describe the operational characteristics of a typical OLT that can be used for GPON links. Consider factors such as interfaces to data switches and CATV equipment, GUI-based network management features, support of switched digital video, size, and electric power requirements.

9.8 Using vendor resources (e.g., such as can be found on the Internet), describe the operational characteristics of a typical ONT that can be used for GPON links. Consider indoor and outdoor units for SDU, MDU, and MTU applications.

List characteristics such as standard telephone and CATV interfaces, high-speed Ethernet connections for enabling computers to access Internet services, size, power, and environmental endurance.

FURTHER READING

1. J. D. Angelopoulos, H.-C. Leligou, T. Argyriou, S. Zontos, E. Ringoot, and T. Van Caenegem, "Efficient transport of packets with QoS in an FSAN-aligned GPON," *IEEE Commun. Mag.*, vol. 42, pp. 92–98, Feb. 2004.
2. Federal Information Processing Standards (FIPS) Publication 197, "Advanced Encryption Standard," Nov. 26, 2001.
3. ITU-T Recommendation G.984.1, "Gigabit-Capable Passive Optical Networks (G-PON): General Characteristics," Mar. 2003.
4. ITU-T Recommendation G.984.2, "Gigabit-Capable Passive Optical Networks (G-PON): Physical Media Dependent (PMD) Layer Specification," Mar. 2003.
5. ITU-T Recommendation G.984.3, "Gigabit-Capable Passive Optical Networks (G-PON): Transmission Convergence Layer Specification," Feb. 2004.
6. ITU-T Recommendation G.984.4, "GPON: ONT Management and Control Interface (OMCI)," June 2004.
7. A. Leon-Garcia and I. Widjaja, *Communication Networks*, McGraw-Hill, Burr Ridge, IL, 2000.
8. B. Meerschman, Y. C. Yi, P. Ossieur, D. Verhulst, J. Bauwelinck, X. Z. Qiu, and J. Vandewege, "Burst bit-error rate calculation for GPON systems," *Proc. Symp. IEEE/LEOS Benelux Chapter*, Enschede, The Netherlands, 2003, pp. 165–168.

10

FTTP CONCEPTS AND APPLICATIONS

When fielding an actual passive optical network, the installation configuration is called by various names and their associated acronyms. A popular general designation is the term *fiber-to-the-premises* (FTTP). In this chapter we define these acronyms and look at some of the fundamental FTTP concepts. In addition, we examine the applications of passive optical networks to various types of neighborhoods, show where optical power splitters can be placed in the network, describe how various clusters of facilities can be connected, examine options for local indoor and outdoor electrical powering solutions, and discuss what current and emerging applications can be expected to run over the network.

10.1 IMPLEMENTATION SCENARIOS

Although the basic concept of a passive optical network is simple, there are many ways of implementing it. In some cases it is possible to run the optical fiber all the way to the customer premises, whereas in other situations it may be more advantageous to terminate the fiber line at some network switching node within or near a neighborhood being served by the PON. Different network design considerations usually are needed, depending on the characteristic of the neighborhood: for example, whether it is a new housing or business development or whether an existing network is being refurbished. In this section we look at some of these issues.

10.1.1 Application Alternatives

The application of PON technology for providing broadband connectivity in the access network to homes, multiple-occupancy units, and small businesses commonly

---

FTTX Concepts and Applications, by Gerd Keiser
Copyright © 2006 John Wiley & Sons, Inc.

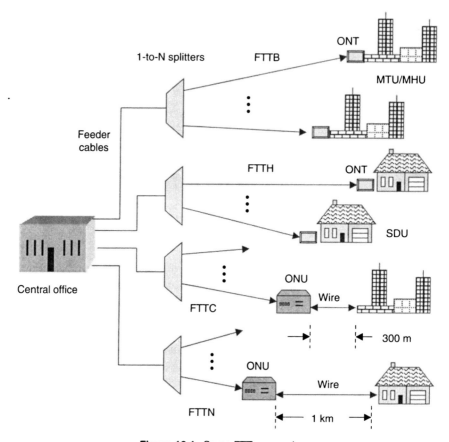

Figure 10.1. Some FTTx scenarios.

is called *fiber-to-the-x*. This application is given the designation FTTx. Here x is a letter indicating how close the fiber endpoint comes to the actual user. Figure 10.1 illustrates some of these scenarios. Among the acronyms shown here and used in the technical and commercial literature are the following:

- FTTB, for *fiber-to-the-business*, refers to the deployment of optical fiber from a central office switch directly into an enterprise.
- FTTC, for *fiber-to-the-curb*, describes running optical fiber cables from central office equipment to a communication switch located within 1000 ft (about 300 m) of a home or enterprise. Coaxial cable, twisted-pair copper wires, optical fiber lines, or some other transmission medium is used to connect the curbside equipment to customers in a building.
- FTTH, for *fiber-to-the-home*, refers to the deployment of optical fiber from a central office switch directly into a home. The difference between FTTB and FTTH is that typically, businesses demand larger bandwidths over a greater part

of the day than do home users. As a result, a network service provider can collect more revenues from FTTB networks and thus recover the installation costs sooner than for FTTH networks.

- FTTN, for *fiber-to-the-neighborhood*, refers to a PON architecture in which optical fiber cables run to within 3000 ft (about 1 km) of homes and businesses being served by the network. FTTN is capable of delivering 20 to 25 Mbps to customers, which allows the simultaneous delivery of services, such as four streams of TV programming, an HDTV channel, and Internet access.
- FTTO, for *fiber-to-the-office*, is analogous to FTTB in that an optical path is provided all the way to the premises of a business customer. As is the case with FTTC, the final transmission segment to the user can include copper-based, optical, or wireless media.
- FTTP, for *fiber-to-the-premises*, has become a popular general term that encompasses the various FTTx concepts. Thus, FTTP architectures include the use of ONUs for FTTC and FTTN applications and the use of ONTs for FTTB and FTTH implementations. An FTTP network can use BPON, EPON, or GPON methodologies.
- FTTU, for *fiber-to-the-user*, is the term used by Alcatel to describe their products for FTTB and FTTH applications.

Of all these acronyms, the commonly used terms are FTTB, FTTH, and FTTP. As described in Section 10.3, an implementation complexity of FTTB and FTTH, as opposed to FTTC or FTTN, is that electric power cannot be supplied to a premises along with the signal on the fiber optic lines. This is a basic difference between using an ONT rather than an ONU. Thus, if the PON fiber terminates at an ONU located some distance from the customer, the copper wire communication line running from the ONU to the premises can provide the electric power needed to operate the end equipment. In contrast, an ONT needs to have its own external electric power supply.

10.1.2 Installation Types

Different design considerations usually are needed for access networks, depending on whether the area being served is a new or an existing neighborhood. As Figure 10.2 illustrates, the three key access network installation concepts are known as greenfield, overbuild, and refurbishment (or access rehabilitation) applications. The term *greenfield* applies to new housing or business park developments. Installing new fiber-based access networks in such areas is very attractive to large service providers, since optical fiber cables or ducts can be placed easily and inexpensively into existing construction trenches throughout the new neighborhood. With the cable or ducts in place, fibers can be run to each new building, thereby providing a 100 percent penetration of a high-capacity network. Although not every homeowner or business may necessarily subscribe to such a network, the chances are that most will. As an additional cost saving to the service provider, since having a high-speed access to each

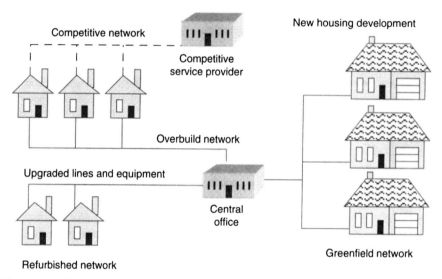

Figure 10.2. Three key access network installation concepts are the greenfield, overbuild, and refurbishment applications.

home or office building is a competitive selling point, the developer may decide to absorb some of the cost of installing the access network.

The term *overbuild* refers to an additional service provider installing a new broadband network capability in a community that already is being served by other competing organizations. For example, a neighborhood often is served by one or more cable operators or digital subscriber line (DSL) providers. Thus, adding another network to such a neighborhood is referred to as a *competitive overbuild*. Typically, the implementing organization will use an overbuild network to deliver more capacity and better service than any of the incumbent service providers can offer. This, of course, pressures the incumbents to upgrade their service offerings to remain competitive.

Refurbishment (sometimes called *rehabilitation*) of a network refers to replacing aging copper wires with optical fibers in a neighborhood. Such an upgrade and expansion of the access network has several advantages. First, it creates new revenue-generating opportunities because of the higher-capacity optical fiber lines. It also greatly reduces network maintenance, since newer more reliable equipment replaces older equipment and transmission lines that may be prone to failures or degradations. Another network term is *overlay*, which refers to a situation where one network runs on top of another. This can enable an increase in services and in capacity of the network without investing in a new or refurbished outside cable plant. For example, dial-up Internet access using modems is an overlay on the traditional telephone network. As another example that earlier chapters illustrate, the use of a 1550-nm wavelength for video services is an overlay to the basic voice and data services sent at 1310 and 1490 nm in a PON. One concept for increasing the capacity of a PON is to add a further WDM overlay (in addition to the current three-wavelength FTTP

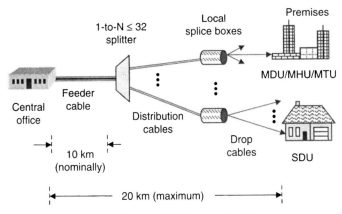

Figure 10.3. General FTTP network architecture.

systems) to create a bandwidth-scalable network to meet future high-bandwidth service demands (see Section 10.5).

10.2 NETWORK ARCHITECTURES

Figure 10.3 shows a general FTTP network architecture. As described in Chapters 1 and 6, the link connecting the central office and the optical splitter is known as a *feeder cable*. The optical splitter may serve up to 32 subscribers. Typically, it is located about 10 km (30,000 ft) from the central office or within 1 km (3000 ft) from the subscribers in a housing neighborhood, a business park, or some other campus. The *distribution cables* originate at the optical splitter. From there they either connect directly to the users or run in a multiple-fiber cable to a local splice box or access terminal. Starting at this splice box, individual *drop cables* connect to the subscribers at the customer premises. In Chapter 12 we describe the implementations of different cable types and illustrate the use of splice boxes, access terminals, and other outdoor enclosures.

10.2.1 Optical Splitter Locations

Two important considerations in designing an FTTP network are the type and the location of the optical power splitter. The choices have impacts on the initial installation cost of a network and on the operational expense and complexity. For example, an FTTP network eventually needs to be modified by adding new users, offering new services, or using more wavelengths for increased capacity. To accommodate these factors when deciding where to place the optical power splitters will require a careful network design analysis. In addition, the decision of where to place the splitter will depend on whether the implementation is in a well-developed neighborhood or in areas that are expanding or undergoing gentrification.

176 FTTP CONCEPTS AND APPLICATIONS

Two possible FTTP network layouts are the centralized and the distributed architectures. The general FTTP network configuration given in Figure 10.3 illustrates the *centralized architecture*. Here there is one central optical power splitter in the network. Typically, there is one 1×32 splitter per FTTP PON, which is housed in an outdoor enclosure or an indoor equipment cabinet if a PON serves a large multiple-occupancy building. A single feeder fiber runs from the OLT in the central office to this splitter. As shown in Figure 10.3, distribution cables containing multiple fibers run from the splitting point to local splice boxes, each fiber being dedicated to a single subscriber endpoint. From the splice box it is easy to run drop fibers for the final connections to the subscribers.

The *distributed architecture* shown in Figure 10.4 has several varieties of splitters located at different strategic places in the network. There are numerous possible configurations with this scheme. For example, at the first power dividing point the optical splitters might range in size from 1×8 to 1×2. Typical secondary splitter sizes near the subscribers might range in size from 1×4 to 1×16. The main motivation for this architecture is to minimize the amount of optical cable that is needed for the distribution and drop fibers, which provides a low initial cost of the outside cable plant.

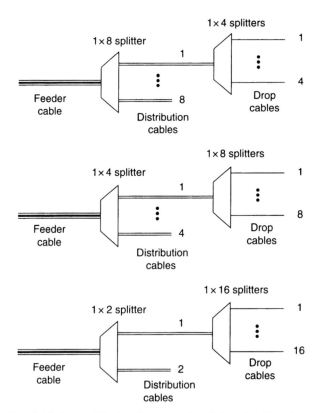

Figure 10.4. The distributed architecture has several varieties of splitters located at different strategic places.

There are several implementation complexities of a distributed architecture compared to a centralized scheme. First, it requires very careful planning of the initial layout of splitter locations and the selection of splitting ratios in order to take into account both current and future subscriber connections. If the selected splitting ratio is too high, there will be many unused splitter ports for drop lines. On the other hand, if the splitting ratio is too low, the local splitters may need to be changed or additional fiber must be installed as more subscribers sign on to the network.

A second factor of a distributed architecture is that it is more expensive to purchase and install several splitters in different locations instead of having centralized splitters in a single enclosure. This also results in additional maintenance and reconfiguration points in the network. When the distributed network requires maintenance (such as determining where a fault has occurred in the network) or if any modifications are needed (such as adding subscribers or rerouting lines), it may be necessary for a service truck to visit multiple locations in the network. Nevertheless, for some sparsely populated areas it may be advantageous to use a distributed-splitter architecture to save the cost of running many long cables to individual scattered clusters of customers.

10.2.2 Network Design Variations

In a fielded installation, a feeder cable will contain multiple fibers for connecting OLTs in the central office to multiple FTTP splitters. As shown in Figure 10.5, these splitters often are located in the same equipment cabinet in order to serve a large number of customers in a neighborhood. Here there are K fibers in the feeder cable, where typically K is larger than the number of PONs being served, which is designated by the letter M in Figure 10.5. This is to allow for additional network growth and to provide extra fibers for other possible independent networks.

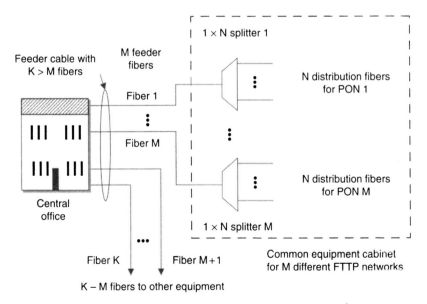

Figure 10.5. Multiple FTTP splitters can be located in the same equipment cabinet.

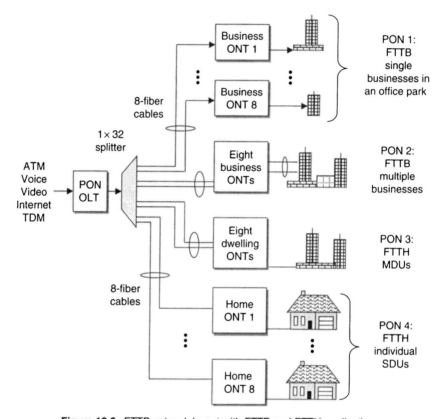

Figure 10.6. FTTP network layout with FTTB and FTTH applications.

As an illustration of a FTTP network layout, Figure 10.6 shows four eight-fiber cables running from the central splitter to local distribution boxes (not shown in the figure). These boxes serve as connection points from which the individual fibers in the cable are attached to drop cables running to homes, apartment buildings, or businesses. Alternatively, some of the fibers in the distribution cable may remain unused until a new customer decides to subscribe to the network.

The four eight-fiber cables shown in Figure 10.6 can be used for different FTTB and FTTH applications on the same PON. Here distribution cable 1 runs to a distribution box in an office park where up to eight separate business buildings can be served. The customers in this FTTB application might have business-related services such as DS1 or ES1 phone and fax lines, Ethernet connections for Internet access, and ATM switches for large data transfers. Distribution cable 2 interfaces to a distribution cabinet located in a large MTU. The ONTs for this FTTB application could be housed in the same wiring room as the distribution cabinet or optical fiber drop cables could connect the cabinet to ONTs located in individual tenant units.

Distribution cable 3 has a function similar to that of distribution cable 2, but here it terminates in an FTTH distribution cabinet located in an MDU. Depending on how the MDU is wired, the ONTs could be in the same location as the distribution cabinet, or individual optical drop fibers could run to apartments in a recently constructed building in which internal optical fibers already may have been installed. In the final scenario, distribution cable 4 serves an FTTH network of individual private homes in a housing development. Here optical drop fibers run from a local splice box to ONTs located at the customer homes.

10.3 LOCAL POWERING OPTIONS

An essential characteristic expected by customers of standard wireline-based telephones is that the service is highly reliable and always available. This means that the telephone service should not be affected by electric power outages, so that emergency calls can be made to fire, police, and hospital personnel. In a standard telephone line the electric power needed to operate a basic phone is provided by means of the copper-based phone line. Therefore, phone service typically is not affected when electric power service from an independently operated utility company is interrupted. Note, however, that many sophisticated modern telephones depend on an external source of electric power beyond that provided by telephone-line powering. This includes cordless phone models that have special built-in features, such as caller identification or a call-answering system.

With an FTTP network, no electric power is going to the customer's premises. Therefore, to have comparable uninterrupted emergency telephony services in a passive fiber optic–based FTTP network requires highly reliable electric powering backup systems at or near the customer's premises. A standard requirement for such uninterrupted systems is the ability to have eight hours of operational backup.

Several options are possible for local powering of the ONT. Each of these has some advantages and limitations, so trade-offs need to be made among factors such as location, deployment logistics, ease of maintenance, cost, reliability, and energy efficiency. The local electric power source for an ONT can be located either inside the customer's premises, next to an individual building, or in a nearby location outside a group of homes or businesses.

10.3.1 Indoor Power Supply

As described in Chapter 6, an ONT can be located next to or inside a customer's premises. Let us first consider the situation where the ONT is inside the premises. There are several advantages to having the powering device indoors. First, installing the power supply is easy and quick since it can be located next to the ONT, and typically, the electric power outlet is close by. There is no need for the service provider to create an expensive and complex powering grid, since FTTP subscribers supply their own electricity. This allows the FTTP network to be deployed rapidly and at a low cost per user, since local power is added simply at the user site whenever that subscriber joins the network.

A typical residential ONT requires a 12-V dc supply and consumes about 15 W of power. To meet the requirement for continuous operation of phone services during emergency situations, the ONT receives its electrical power from an indoor

TABLE 10.1 General Specifications of an Indoor UPS

Characteristic	Requirement
Voltage	12 V
Power	15 W (minimum)
Power connection	110/220 V external supply; secured plug attachment to an indoor power source
Status monitoring signals	Battery on Low battery Replace battery Battery missing
Battery issues	Seven ampere-hour capacity; easily accessible and replaceable by the customer

uninterruptible power supply (UPS). In case of a power outage, the ONT will disable services that are not critical, such as data and video, in order to reduce power consumption. In such an operational state an ONT consumes about 7 W of power. The UPS can meet the requirement of eight hours of operation through a standard internal battery that has a 7 ampere-hour capacity.

Table 10.1 lists some specifications of an indoor UPS. Since such a unit is located inside a building, its electronics and enclosure do not need to be environmentally rugged. Thus, an indoor UPS is much less expensive and more reliable than an outdoor supply, which is exposed to harsher environments. However, to have a reliable overall system, the ONT needs to send supply alarms and management data to notify the FTTP network operational personnel of any anomalous supply behavior or of the battery condition.

One important parameter is the status of the internal backup battery. When it is nearing the end of its life, the OLT should receive a signal indicating that the battery needs to be replaced. A logistics problem with this scheme relates to whether the subscriber or the service provider should be responsible for replacing the battery. If the service provider has the task to do this, there is the expense of sending a service technician, and that person needs to have access to the customer's premises. In the other case, the customer must take the initiative to obtain and replace the battery.

10.3.2 Outdoor Power Supply

Now consider the situation where the ONT and the power supply are on the outside of a building. Similar to an indoor installation, the FTTP network can be deployed fairly rapidly, since local power is added simply at the user site whenever that subscriber joins the network. However, in this case, attaching the UPS to the electric power source may be slower for existing premises. That is because typically an electric outlet is not readily available outdoors. In addition, the power line running to the UPS must be protected and meet outdoor safety requirements. For greenfield developments this type of electric connection can be made available through preplanning before the buildings are constructed.

TABLE 10.2 General Specifications of an Outdoor UPS

Characteristic	Requirement
Voltage	12 V
Power	15 W (minimum)
Power connection	110/220 V external supply; secure attachment to an outdoor power source
Status monitoring signals	Battery on
	Low battery
	Replace battery
	Battery missing
Battery issues	Seven-ampere-hour capacity
	Easily accessible and replaceable by a service technician
	Battery may need to endure very cold environments

The electronic components and the enclosure of any equipment located outdoors need to be environmentally hardened to protect them from temperature extremes, moisture, and intrusion attempts. This makes it more expensive than comparable indoor equipment. In addition, the battery life typically is shorter than for an indoor application. However, it is easier to service an outdoor unit and to replace the battery, since access to the customer's premises is not required. Table 10.2 lists some specifications of an outdoor UPS.

10.3.3 Network Powering

A third alternative to providing electric power to an ONT is to install a large UPS outside on a pole or in an outdoor cabinet. Such equipment would be attached to the public electric lines and would be able to supply electric power over twisted-pair copper wires to the ONTs in multiple homes or businesses. To provide an efficient power transfer to the customer's ONT, a 48-V dc UPS typically is used. Since most ONTs run on 12 V dc, this method requires a step-down transformer at each ONT.

The disadvantages of such UPS solutions include the need to secure a contract with the electric power company, the requirements for step-down transformers, high costs for the large complex UPS equipment, the need to run copper wire to each premises, and the need for a large outdoor enclosure. In addition, if few customers sign on to the network, the capacity of a large multiple-premises UPS is underused.

10.4 SERVICE APPLICATIONS

Traditionally, subscribers use separate networks for connecting to services such as telephony, video broadcasting, and the Internet. For example, they might be connected independently by means of a public switched telephone network from service provider A, a cable network for video and TV viewing, and a digital subscriber line (DSL) from service provider *B* for Internet access. A triple-play FTTP scheme can integrate these three standard services onto a single network using different wavelengths.

10.4.1 Bandwidth Requirements

The challenge for service providers is how to offer emerging and expanding applications that require large amounts of bandwidth. These include IP telephony, video on demand (VOD), multiple time-shifted high-definition television (HDTV) channels, interactive entertainment games, two-way videoconferencing, externally monitored home security, and streaming of multimedia Internet content. Most of these services are used directly by the consumer. Home security monitoring can be performed by a distantly located organization that has control over indoor and outdoor security cameras, fire alarms, and intrusion detection systems that monitor the user's premises.

Table 10.3 lists the maximum bandwidth requirements per subscriber for some of these types of services for a home. Obviously, such high bandwidth use means that there are several active users at each residential site with access to multiple HDTV sets and high-performance personal computers. Small businesses typically have similar bandwidth requirements, but with a different mix of service needs. Thus, service providers need to able to offer bandwidths of 75 to 100 Mbps per premises for existing and emerging FTTP network applications.

10.4.2 Video Service Issues

One way that a customer can receive video services in an FTTP network is by means of a digitized and compressed bit stream along with the voice and data. This method commonly is called *video over IP*. The other means of receiving video is through the use of a separate *overlay* wavelength. The categories of video transmission include video broadcasting, video on demand, and video conferencing. The first two of these are one-way transmissions but may have a low-rate feedback link for video file requests. Videoconferencing uses a full-duplex interactive link.

TABLE 10.3 Maximum Bandwidth Requirements per Premises for Various FTTH Services

Service	Bandwidth Needed (Mbps)
Two HDTV channels (20 Mbps each)	40
Two standard digital TV channels (3 Mbps each)	6
Two Internet users (2 Mbps each)	4
Standard telephone	0.1
Video phone	2
Home security monitoring	2
Total	54.1

The *video broadcasting* method is used to send video content to one or more passive viewers who have no control over the session. Its applications include entertainment, distribution of corporate training material, technical or business presentations, continuing education material or lectures, and speeches. *Video on demand* offers viewers the interactive options to start, stop, rewind, or advance the video content. Typically, it is used for viewing stored video files. Among its applications are education, entertainment, marketing, and training.

Videoconferencing uses both audio and video transmissions to enable people in two different locations to participate in a real-time face-to-face meeting. Since the transmission is full-duplex, videoconferencing is not possible with the standard FTTP video overlay wavelength, since that is a one-way transmission. However, the H.323 videoconferencing standard allows an IP-based conference participation (e.g., at 384 kbps) that can be initiated from any personal computer attached to the network. Among its many applications are business communications, telemedicine, training, customer services, and the use of an electronic whiteboard.

Table 10.3 shows that among the video services an MPEG2-encoded HDTV channel demands the most bandwidth in an FTTP network. If one user on each of the 32 branches of an FTTP network desires to view such a 20-Mbps HDTV channel, there is not enough bandwidth available on the 1490-nm downstream wavelength. This drives the need for using the separate 1550-nm wavelength for this application.

To accommodate the various video services, both a broadcast and a narrowcast method are used. As Figure 10.7 shows, the video-related equipment in the central office consists of a headend transmitter that receives analog broadcast TV signals from sources such as a TV broadcast studio or a satellite link. Following the transmitter is an optical amplifier that boosts the optical signal level before it is sent out over the PON feeder cable. This amplification is necessary, since to have a clear picture the optical power level of the video signal at the video receiver at the customer's premises must be at least -5 dBm. Figure 10.8 shows two configurations of a compact erbium-doped fiber amplifier (EDFA) that are suitable for such applications.

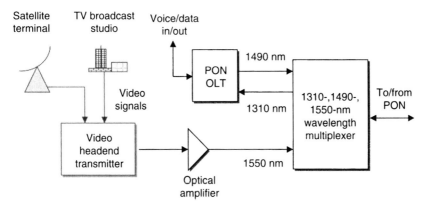

Figure 10.7. Interfaces to video-related equipment in the central office.

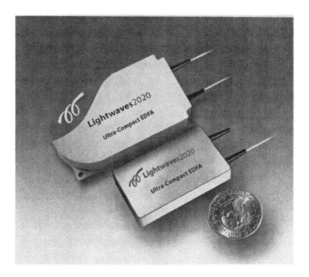

Figure 10.8. Two package sizes for compact optical amplifiers. (Photo courtesy of Lightwaves-2020; www.lightwaves2020.com.)

Note that in contrast to DWDM applications, in which an optical amplifier must operate over spectral ranges on the order of 30 nm, in a standard FTTP network the EDFA needs to amplify only a single 1550-nm wavelength.

10.5 EXPANDED WDM PON

As is the case with any technology, there are ample opportunities for enhancements of FTTP network architectures. These include adding many more wavelengths, increasing the transmission speed per wavelength, and extending the reach of the network. In this section we examine FTTP network enhancements using wavelength-division multiplexing.

The ever-increasing demand for higher-capacity triple-play services can quickly lead to demands of close to 100 Mbps per subscriber. Since a standard three-wavelength FTTP network will not be able to satisfy such demands, a possible enhancement is to use more wavelengths to create a WDM-PON. The idea in this method is to use a separate wavelength for each transmitting ONT, so that an ONT can send its information continuously over the shared upstream fiber without having to wait for a specific assigned transmission time slot.

In such a WDM PON it is desirable to have a *colorless* ONT, which means that no ONT should be assigned a fixed transmission wavelength. An obvious but extremely expensive solution is to use a tunable laser at each ONT. Since the low cost of end equipment is a driving factor in PON implementations, this is not a feasible solution. Thus, a major challenge for a WDM PON implementation is to have a low-cost high-output optical source.

One method that has been proposed is to use spectral splicing of a single-broadband relatively inexpensive light source. Various techniques are being explored to achieve this. One idea is to use a wavelength-locked Fabry–Perot (FP) laser diode, as described by Lee in the OFC 2005 paper. This technique uses narrow spectral slices from a broadband light source to force the FP laser to operate in a quasi singlemode, whereby the mode partition noise of the FP laser diode is suppressed sufficiently to allow the device to be used as a WDM source.

As other WDM PON techniques are being developed, the interested reader is referred to the recent literature. In addition, descriptions of new technology will be posted periodically on the book Web site (see page xvi).

10.6 SUMMARY

The application of PON technology for providing broadband connectivity in the access network to homes, multiple-occupancy units, and small businesses commonly is called fiber-to-the-x. This application is given the designation FTTx. The letter x indicates how close the fiber endpoint comes to the actual user. The most commonly used acronyms are FTTB (fiber-to-the-business), FTTH (fiber-to-the-home), and FTTP (fiber-to-the-premises).

Different design approaches usually are needed for access networks, depending on whether the area being served is a new or an existing neighborhood. The three key access network installation concepts are known as greenfield, overbuild, and refurbishment (or access rehabilitation) applications. The term *greenfield* applies to new housing or business park developments. The term *overbuild* describes the situation where another service provider installs a new broadband network capability in a community that already is being served by other competing organizations. *Refurbishment* (sometimes called *rehabilitation*) of a network refers to replacing aging copper wires with optical fibers in a neighborhood.

In a general FTTP network architecture, the link connecting the central office and the optical splitter is known as a feeder cable. The optical splitter may serve up to 32 subscribers. Usually, it is located about 10 km (30,000 ft) from the central office or within 1 km (3000 ft) from the subscribers in a housing neighborhood, business park, or other campus. The distribution cables originate at the optical splitter. From there individual fibers either connect directly to the users or run in a multiple-fiber cable to a local splice box. Starting at this splice box, individual drop cables connect to the subscriber's premises.

With an FTTP network there is no electric power going to the customer's premises. Therefore, to have uninterrupted emergency telephony services in a passive fiber optic–based FTTP network requires highly reliable electric powering backup systems at or near the customer's premises. A standard requirement for such uninterrupted systems is the ability to have eight hours of operational backup. The electric power source for an ONT can be located either inside the customer's premises, next to an individual building, or in a nearby location outside a group of homes or businesses. Each option has advantages and limitations, so trade-offs need to be made among factors such as location, deployment logistics, ease of maintenance, cost, reliability, and energy efficiency.

PROBLEMS

10.1 Consider the geographical distribution of 32 homes and businesses shown in Figure 10.9. The premises are grouped into clusters of 8 and 16. These clusters are 15 km from the central office and are separated by 4 km. Show how these 32 locations might be connected with a FTTP network that uses a centralized slitter and with distributed-splitter FTTP network architecture.

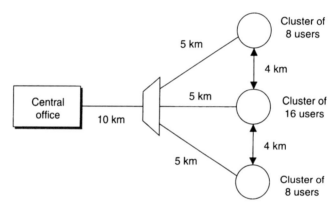

Figure 10.9. Geographical distribution of 32 homes and businesses (for Problems 10.1 and 10.2).

10.2 When deciding on whether to use a centralized or a distributed architecture for the FTTP network described in Figure 10.9, what are some issues related to fiber cable price, installation costs of cable and splitters, expenses of enclosures for splitters, and network reconfiguration or expansion flexibility?

10.3 Consider a small industrial park containing 64 companies housed in separate buildings. Suppose that the park is a 2 × 2 km square and is located 10 km from a central office. Describe the advantages and limitations of using FTTN, FTTC, and FTTP deployments for providing broadband access to these businesses.

10.4 An ONT intended for indoor installation requires a 12-V dc power supply and consumes 15 W of electric power during normal operation. If there is an electric power outage, the ONT disables nonessential services and now consumes 7 W of power, supplied by backup batteries. Using vendor data sheets, select an indoor uninterruptible power supply (UPS) that satisfies these requirements.

10.5 An ONT intended for outdoor installation requires a 12-V dc power supply and consumes 15 W of electric power during normal operation. If there is an electric power outage, the ONT disables nonessential services and now consumes 7 W of power, supplied by backup batteries. Using vendor data sheets, select an outdoor uninterruptible power supply (UPS) that satisfies these requirements. Assume that the UPS must operate over temperature extremes of $-25°C$ to $+35°C$.

10.6 Draw a wiring diagram of how an outdoor UPS for powering an outdoor ONT might be connected to the building power source. What special safety precautions must be considered in this application?

10.7 List and describe the performance characteristics of a compact EDFA, such as that shown in Figure 10.8, that can be used for boosting the 1550-nm video signal at the central office to an appropriate level in order to have clear video reception at a customer site.

10.8 Describe how a video conferencing system can be implemented over FTTP networks using standard techniques such as prescribed by H.323. Which PON technology (BPON, EPON, or GPON) can best handle the bandwidth requirements?

FURTHER READING

1. M. Abrams, P. C. Becker, Y. Fujimoto, V. O'Byrne, and D. Piehler, "FTTP deployments in the United States and Japan: equipment choices and service provider imperatives," *J. Lightwave Technol.*, vol. 23, pp. 236–246, Jan. 2005.
2. N. J. Frigo, P. P. Iannone, and K. C. Reichmann, "A view of fiber to the home economics," *IEEE Commun. Mag.*, vol. 42, pp. S16–S23, Aug. 2004.
3. P. E. Green, "Fiber to the home: the next big broadband thing," *IEEE Commun. Mag.*, vol. 42, pp. 100–106, Sept. 2004.
4. C. H. Lee, "Passive optical networks for FTTx applications," *OFC 2005 Conf. Proc.*, Anaheim, CA, Mar. 2005, Paper OWP3.
5. M. Nakamura, H. Ueda, S. Makino, T. Yokotani, and K. Oshima, "Proposal of networking by PON technologies for full and Ethernet services in FTTx," *J. Lightwave Technol.*, vol. 22, pp. 2631–2640, Nov. 2004.
6. S.-J. Park, C.-H. Lee, K.-T. Jeong, H.-J. Park, J.-G. Ahn, and K.-H. Song, "Fiber-to-the-home services based on wavelength-division-multiplexing passive optical network," "*J. Lightwave Technol.*, vol. 22, pp. 2582–2591, Nov. 2004.
7. D. J. Shin, D. K. Jong, H. K. Shin, J. W. Kwon, S. Hwang, Y. Oh, and C. Shim, "Hybrid WDM/TDM-PON with wavelength-selection-free transmitters," *J. Lightwave Technol.*, vol. 23, pp. 187–195, Jan. 2005.

11

FTTP NETWORK DESIGN

In this chapter we describe how individual passive optical components, transceivers, and fibers are put together to form a complete FTTP network. In Section 11.1 we look at the design criteria for an end-to-end link. This includes examining the components that are available for a particular application and seeing how they relate to the system performance criteria (such as bit error rate and dispersion). In addition, certain operational factors in a link may contribute to signal impairment, which results in a reduction of the ideal signal-to-noise ratio. In Section 11.2 we then show that for a given set of components and a specified set of system requirements, one can carry out an optical power-budget analysis to determine whether the fiber optic link design meets the attenuation allowed in both directions across the FTTP network. To assist in such FTTP network design calculations, in Section 11.3 we introduce the use of software-based photonic design automation tools that can be run on a personal computer. Such tools employ a graphical-programming language that allows the modeling and simulation of lightwave components, links, and networks. Further information on such tools, up-to-date interactive simulation examples, and links to Web sites of simulation tool developers may be found on the book Web site (see page xvi).

How to perform a system rise-time analysis to verify that the overall bandwidth requirements or link capacities are met is described in Section 11.4. For most FTTP networks the rise-time criteria are well satisfied, since the transmission distances are relatively short. Finally, in Section 11.5 we examine various link and equipment redundancies that ensure the high network reliability and uptimes that customers expect and demand to have.

11.1 DESIGN CRITERIA

The design of high-quality links for FTTP optical access networks involves a series of trade-offs among the many interrelated performance variables of each

---

FTTX Concepts and Applications, by Gerd Keiser
Copyright © 2006 John Wiley & Sons, Inc.

component based on the system operating requirements. Thus, the actual link design analyses may require several iterations before they are completed satisfactorily. Since performance and cost constraints are very important factors in a transmission link, the designer must choose the components carefully to ensure that the link meets the operational specifications over the expected system lifetime without overstating the component requirements.

To assist in these trade-off calculations, network designers can use software-based modeling and simulation tools that can be run on a personal computer. Such tools include an extensive library of standard predefined components that enable the performance of many different network layout scenarios to be simulated. Section 11.3 has more details on these types of tools.

11.1.1 System Requirements

The following key system requirements are needed in analyzing a link:

- The transmission distance
- The number and types of optical splitters
- The data rate or channel bandwidth
- The bit error rate (BER)
- The number of wavelength channels
- The available optical attenuation budget
- The desired optical power margin for the link
- Power penalties due to various system impairments

To fulfill these requirements the FTTP network designer has a choice of many passive and active components. Table 11.1 lists some of these and gives their

TABLE 11.1 Types and Characteristics of Components Used for Optical Link Design

Component	Type	Characteristics or Use
Optical fibers	Single-mode or multimode	Attenuation, dispersion, SBS tolerance
Optical cables	Aerial, duct, or underground	Fiber count, strength members, FTTP segment application
Light source	LED, DFB laser, or FP laser	Modulation rate, output power, wavelength, spectral width, cost
Photodetectors	pin or APD	Sensitivity, responsivity
Connectors	Single or multiple channel	Loss, size, mounting type
Optical splitters	Fiber-based or PLC	Size ($1 \times N$), insertion loss, packaging
Passive components	Optical filters, optical isolators, and power couplers	Wavelength response, loss, size, cost, reliability
Transceivers (OLT, ONT, or ONU)	Indoor or outdoor	Environmental ruggedness, size, cost, reliability, electric power use

TABLE 11.2 Characteristics of Related Equipment and Materials for FTTP Link Design

Component	Type	Characteristics or Use
Test instruments	BER tester, OTDR, spectrum analyzer, power meter	Optical power, wavelength range, portability, cost, reliability
Cabinets	Aerial, aboveground, or underground; indoor or outdoor	Environmental ruggedness, size, cost, reliability
Power supplies	Indoor or outdoor	Size, cost, reliability, backup battery
Cable ducts	Macro- or mini-duct	Material, size, cost, reliability

characteristics or notes how they are used. In addition, to have a comprehensive system design an engineer also needs to consider related equipment and materials, such as electronics cabinets, power supplies, and test instruments. Table 11.2 lists some of these and their characteristics. How and where such equipment and materials are used are discussed further in Chapters 12 and 13.

11.1.2 System Margin

System margin (also called *loss margin*) is an optical-power safety factor for link design. This involves adding extra decibels to the power requirements to compensate for possible unforeseen link degradations. These degradations could arise from factors such as a dimming of the light source over time, losses resulting from aging of other components in the link, the possibility that certain splices or connectors in the actual link have a higher than anticipated loss, or additional losses that may occur when a cable is repaired.

ITU-T Recommendation G.957 specifies that a system margin ranging from 3.0 to 4.8 dB should be allowed between the transmitter and the receiver to offset possible equipment degradation. In a conventional telecommunication system design, engineers typically added a system margin of 3 to 10 dB, depending on the performance requirements of the application, the number of possible repairs, and the system cost. Improvements in design techniques, the use of software-based modeling and simulation tools, and the drive to optimize network deployment costs has resulted in considerable shrinkage of this budget margin. Currently, an accepted standard design rule in the industry is that budget-loss margins should not be greater than 3 dB. Therefore, tight margins are used for PONs, since the distances are relatively short and there are fewer active components than in metro or long-haul networks. An appropriate choice might be few decibels of margin.

11.1.3 Power Penalties

Certain operational factors in a link usually contribute to signal impairment. Among these factors are modal noise, chromatic dispersion, polarization-mode dispersion, reflection noise in the link, low extinction ratios in the laser, or frequency chirping. When any of these dispersion or nonlinear effects contributes to signal impairment, there is a reduction in the signal-to-noise ratio (SNR) of the system from the ideal case. This reduction in SNR is known as the *power penalty* for that effect, which generally is expressed in decibels.

Let us look at the details of some of these power penalties.

1. A signal pulse from a light source contains optical power from a certain slice of wavelength spectrum. For example, a modulated laser diode source may emit pulses that have a 0.1-nm spectral width. *Chromatic dispersion* originates from the fact that each wavelength travels at a slightly different velocity in a fiber. The power penalty arising from chromatic dispersion can be calculated from

$$P_{CD} = -5 \log [1 - (4BLD_{CD} \Delta\lambda)^2] \tag{11.1}$$

where B is the bit rate in Gbps, L is the fiber length in kilometers, D_{CD} is the chromatic dispersion in ns/(nm·km), and $\Delta\lambda$ is the spectral width of the source in nanometers. Whether one implements high-speed single-wavelength or WDM networks, this effect can be mitigated by the use of various dispersion compensation schemes. To keep the power penalty less than 0.5 dB, a well-designed system should have the quantity $BL D_{CD} \Delta\lambda < 0.1$. This effect is not a major concern for PON designs that are based on ITU-T Recommendations G.983, G.984, and G.985, because of the relatively short distances and the data rates, which are less than 2.5 Gbps. However, it needs to be considered more carefully in future PON upgrades to higher data rates or if there are DWDM overlays.

2. *Polarization-mode dispersion* (PMD) arises in single-mode fibers because the two fundamental orthogonal polarization modes in a fiber travel at slightly different speeds, owing to fiber birefringence (see Section 2.5). Since this is a time-varying effect, it has been estimated that to avoid having a power penalty of 1 dB or greater for a fractional time of 30 minutes per year, the average differential time delay between the two different polarization states must be less than 0.14 of the bit period. In general, this effect is negligible for the 20-km distances and the data rates of less than 10 Gbps, which characterize standard FTTP applications.

3. The *extinction ratio* (ER) in a laser is defined as the ratio of the *on* power for a logic 1 to the *off* power for a logic 0. Ideally, one would like the extinction ratio to be infinite, so that there would be no power penalty from this condition. However, the extinction ratio must be finite during a transmission time period. This is necessary to reduce the rise time of laser pulses so that signal distortions do not occur. The power penalty increases significantly for lower extinction ratios. If r_e represents the ratio of the average power in a logic 1 to the average power in a logic 0, the power penalty P_{ER} arising from a finite extinction ratio is given by

$$P_{ER} = -10 \log \frac{r_e - 1}{r_e + 1} \tag{11.2}$$

In practice, optical transmitters have minimum extinction ratios ranging from 7 to 10 (8.5 to 10 dB), for which the power penalties range from 1.25 to 0.87 dB.

A minimum extinction ratio of 18 is needed to have a power penalty of less than 0.5 dB. Note that as described in Section 4.5, for PON applications the laser must be turned off almost completely when it is not transmitting. This is done to avoid significant raises in the OLT noise floor, as would happen if all 32 ONT lasers were biased slightly on at all times. However, during a transmission burst the laser must have an appropriate extinction ratio to avoid signal distortions.

4. A power penalty due to *frequency chirping* arises from the fact that the light output from an optical source experiences a dynamic spectral broadening (or a *frequency chirp*) when the laser is modulated directly. The chirping power penalty is reduced for higher bias settings, but this increases the penalty arising from the lower extinction ratio. When analyzed in conjunction with the extinction ratio degradation, the combined power penalty typically is less than 2 dB for an extinction ratio setting of about 10.

5. As described in Section 2.6, nonlinear effects occur when there are high light power densities (optical power per cross-sectional area) in a fiber. Since the wavelengths used in an FTTP network are widely separated, the two nonlinear effects of concern are *stimulated Brillouin scattering* (SBS) and *stimulated Raman scattering* (SRS). SBS produces a scattered wave that propagates principally in the backward direction in single-mode fibers. This backscattered light experiences gain from the forward-propagating signals, which leads to depletion of the signal power. This is especially undesirable in analog video transmission, since the depletion process causes noticeable signal degradation. The signal depletion becomes increasingly stronger as the optical power level increases, until a threshold of about 17 dBm (50 mW) is reached in standard single-mode fibers. At this point any additional launch power gets absorbed by the backscattered power. The SRS process generates scattered light at a wavelength longer than that of the incident light. If another signal is present at this longer wavelength, the SRS light will amplify it and the pump-wavelength signal will decrease in power.

11.2 LINK POWER BUDGET

In carrying out a *link power-budget analysis*, one first needs to determine the link *power margin* or allowed *attenuation range*. Simply, this is the maximum allowed difference in light power level between the optical transmitter output and the minimum receiver sensitivity needed to establish a specified BER. For FTTP power-budget calculations, the ITU-T has defined the three attenuation range classes listed in Section 6.6.

As shown in Figure 11.1, the attenuation that is allowed can be allocated to fiber, splitter, splice, and connector losses, plus any additional optical power losses that may arise from other components, possible device degradations, transmission-line impairments, or temperature effects. If the choice of components did not allow the desired transmission distance to be achieved for the FTTP network, the components might have to be changed.

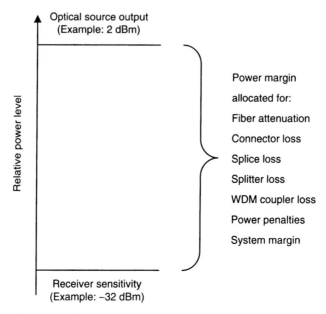

Figure 11.1. Power margin allocation to various link loss factors.

11.2.1 Power-Budgeting Process

The first step in calculating a power budget is to decide at which wavelength to transmit and then select components that operate in this region. For FTTP networks the main three wavelengths are 1310, 1490, and 1550 nm. A link power-budget calculation needs to be done at all three of these wavelengths. The two main reasons for doing this is that the fiber attenuation varies with wavelength and the receiver sensitivities are different in the upstream and downstream directions since the transmission rates are not the same. As shown in Figure 11.2, the attenuation decreases with increasing wavelength. For example, the nominal attenuation of 0.4 or 0.5 dB/km at 1310 nm is about twice the 0.25-dB/km nominal attenuation at 1550 nm.

Once a wavelength has been selected for doing a power-budget calculation, the next step is to correlate the system performances of the major optical link building blocks. For an FTTP network these include the optical receiver, transmitter, splitter, connectors, splices, and fiber. Normally, the designer chooses the characteristics of two of these elements, selects average performance characteristics of the other components, and then does a power-budget calculation to see if the link margin requirements are met. A typical procedure is first to select the photodetector and the optical source. That is, the design engineer selects which class of optics the link should adhere to (see Section 6.6). Then the designer calculates how far signals can travel over a particular fiber with specific optical splitters, connectors, and splices.

Figure 11.3 shows a hypothetical FTTP link between an OLT and an ONT. Here there are four connectors within the link, an optical power splitter in the central region, N splices located along the cable path, and WDM couplers at the OLT and at

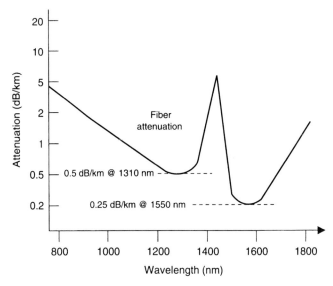

Figure 11.2. Optical fiber attenuation as a function of wavelength.

the ONT. The optical power arriving at the photodetector depends on the amount of light coupled into the fiber minus the losses incurred along the path. The link-loss budget is derived from the sequential loss contributions of each element in the link. Each of these losses is expressed in decibels as

$$\text{loss} = 10 \log \frac{P_{out}}{P_{in}} \quad (11.3)$$

where P_{in} and P_{out} are the optical powers entering and exiting, respectively, a fiber, splice, connector, or other link element.

The link-loss budget simply considers the total optical power loss P_T that is allowed between the light source and the photodetector, and allocates this loss to factors such as cable attenuation, connector and splice losses, losses in other link components, and system margin. Thus, referring to Figure 11.3, if P_S is the optical

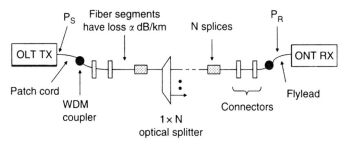

Figure 11.3. Losses in a hypothetical FTTP link between an OLT and an ONT.

power from the source entering the WDM coupler at the OLT and if P_R is the minimum receiver sensitivity needed for a specific BER at the ONT, then

$$P_T = P_S - P_R$$
$$= 4 \times \text{connector loss} + \alpha L + N \times \text{splice loss} + \text{splitter loss}$$
$$+ \text{WDM coupler losses} + \text{system margin} \quad (11.4)$$

where α is the fiber attenuation in dB/km and L is the link length. As noted in Section 11.1.2, the system margin can be selected to be around 3 dB.

Now let us look at some examples of how to calculate a link-loss budget. We use examples of links operating at 1310 and 1490 nm.

11.2.2 FTTP 1310-nm Power Budget

An engineer plans to design a BPON-based FTTP network with a maximum link length of 20 km, as shown in Figure 11.4. The link design will use a class B attenuation range. This will allow a maximum attenuation of 25 dB across the optical data network (ODN) between its interfaces with the two WDM couplers. The network will use the following components and conditions:

- A 1×32 optical power splitter that has a throughput loss of 16.5 dB is used.
- An installed fiber that meets the G.652 specification is used. At 1310 nm the fiber attenuation is 0.4 dB/km and at 1490 nm it is 0.25 dB/km.

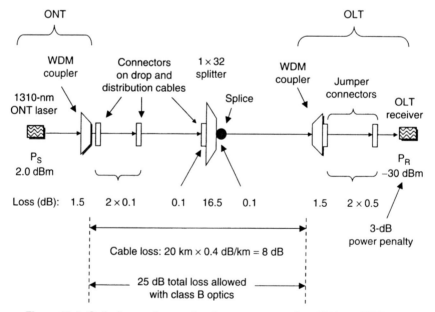

Figure 11.4. Optical power losses of various components in a 1310-nm FTTP link.

- For the 20-km cable span between the two WDM couplers there are a maximum of three low-loss LC connectors and one fusion splice with a loss of 0.1 dB each.
- There is a wavelength-combining and a wavelength-separating component at both the OLT and the ONT. Assume that each of these WDM components has a throughput loss of 1.5 dB.
- The downstream data rate is 1.25 Gbps and the upstream rate is 622 Mbps. For this system we assume that the BER desired is 10^{-11} (i.e., at most one error can occur for every 10^{11} bits sent).
- There is a 3-dB power penalty associated with the optical receiver at the OLT, since it needs to operate in a burst mode.

Note that these loss values are conservative. Typically, in a practical fielded FTTP network some of these losses are larger, and there may be more connectors and splices, which also increases the link loss. In addition, the effects of other power penalties besides that due to the burst-mode receiver have not been included. See Problem 11.6 for possible effects of including other losses.

The first step is to calculate the 1310-nm power budget. The engineer selects an ONT containing a laser diode that can launch $P_S = 2.0$ dBm (1.6 mW) of 1310-nm optical power into the fiber and an OLT that uses an InGaAs pin photodiode. The photodetector has a sensitivity of $P_R = -30$ dBm at 622 Mbps for a 10^{-11} BER at 1310 nm. The total optical power loss allowed between the ONT laser and the OLT photodetector can than be at most $P_T = P_S - P_R = 32$ dB.

Assume that here, because of the way the equipment is arranged, a short optical jumper cable is needed in the central office between the incoming feeder cable and the OLT equipment rack. Assume that the jumper cable introduces a loss of 1.0 dB, which is due mainly to a 0.5-dB mated-connector loss at each end of the jumper.

Two low-loss LC connectors (0.1 dB per mated pair) are located at each end of the drop cable running from the breakout box to the ONT at the subscriber's premises. There is another LC connector at the interface between the splitter and the distribution cable. The main splice is at the end of the feeder cable where it attaches to the optical splitter.

A convenient procedure for calculating the power budget is to use a tabular or spreadsheet form. This is illustrated in Table 11.3, which gives the spreadsheet for calculating the 1310-nm link power budget. This table lists the components in the leftmost column and the associated optical output, sensitivity, or loss in the center column. The rightmost column gives the power margin available after successively subtracting the component loss from the total optical power loss that is allowed between the light source and the photodetector. In this case the allowable loss between the two WDM couplers is 25 dB for class B operation. The final system margin is 0.1 dB, which is a very tight safety factor for this link. Therefore, operation at 1310 nm is barely adequate in this case. Thus, the designer may need to consider the use of class C components, which adds 5 dB to the attenuation range. Alternatively, the splitter size or link length can be reduced.

TABLE 11.3 Spreadsheet for Calculating the 1310-nm BPON-Based FTTP Link Power Budget for a Class B Attenuation Range

Component/Loss Parameter	Output/Sensitivity/Loss	Power Margin (dB)
ONT laser diode output	2.0 dBm	
OLT detector sensitivity at 622 Mbps	−30 dBm	
Allowed loss [2−(−30)]		32.0
WDM coupler loss (2 × 1.5 dB)	−3.0 dB	29.0
Central-office patch cord loss	−1.0 dB	28.0
OLT receiver power penalty	−3.0 dB	25.0
Power available for class B link		25.0
Power splitter loss (1 × 32)	−16.5	8.5
Splice loss (1 × 0.1 dB)	−0.1	8.4
Connector loss (3 × 0.1 dB)	−0.3	8.1
Cable attenuation (20 km × 0.40 dB/km)	−8.0 dB	0.1 (system margin)

11.2.3 FTTP 1490-nm Power Budget

Transmission of digitized information at the other end of the FTTP network spectrum is done at 1490 nm. Three parameter values change here, as shown in Figure 11.5. First, the optical fiber attenuation drops to 0.25 dB/km. Second, a BPON-based FTTP network uses a higher data rate downstream at 1490 nm, so the pin photodiode receiver sensitivity is $P_R = -26$ dBm for this wavelength at 1.25 Gbps to maintain a 10^{-11} BER. Assume that the laser diode transmitters have fiber-coupled outputs of $P_T = 3$ dBm

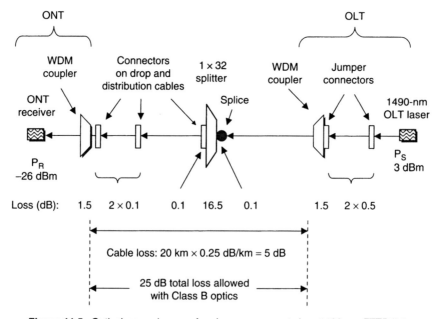

Figure 11.5. Optical power losses of various components in a 1490-nm FTTP link.

TABLE 11.4 Spreadsheet for Calculating the 1490-nm BPON-Based FTTP Link Power Budget for a Class B Attenuation Range

Component/Loss Parameter	Output/Sensitivity/Loss	Power Margin (dB)
OLT laser diode output	3.0 dBm	
ONT detector sensitivity at 622 Mbps	−26 dBm	
Allowed loss [3−(−26)]		29.0
WDM coupler loss (2 × 1.5 dB)	−3.0 dB	26.0
Central office patch cord loss	−1.0 dB	25.0
Power available for class B link		25.0
Power splitter loss (1 ×32)	−16.5	8.5
Splice loss (1 × 0.1 dB)	−0.1	8.4
Connector loss (3 × 0.1 dB)	−0.3	8.1
Cable attenuation (20 km × 0.25 dB/km)	−5.0 dB	3.1 (system margin)

(2 mW). The other parameter values are the same as for the 1310-nm case. The only exception is that now there is no burst-mode power penalty for the ONT receiver since it operates in a conventional mode. The various power losses are listed in Table 11.4. The system margin is 3.1 dB, which is acceptable for a 20-km link.

11.3 PHOTONIC DESIGN AUTOMATION TOOLS

Software-based photonic design automation tools that can be run on a personal computer make the network design process more efficient, less expensive, and faster. Such tools employ a graphical-programming language that allows integrated modeling and simulation of lightwave components, links, and networks. In this section we first give some highlights of these tools and then give an example of a modeling program for FTTP networks from RSoft Design Group. Further information on such tools, up-to-date interactive simulation examples, and links to Web sites of simulation tool developers may be found on the book Web site (see page xvi).

11.3.1 Modeling Tool Characteristics

To allow a design engineer to visualize and simulate a network quickly, the modeling and simulation programs normally have the following features:

- They have a library of graphical icons and interconnection tools that enable an engineer to create a network schematic using a graphical user interface (GUI). The icons represent system components (e.g., optical fibers, power splitters, amplifiers, laser transmitters, and receivers) and measurement instruments (e.g., signal generators, optical power meters, spectrum analyzers, and optical-loss sets).
- They allow a user to interact easily with the program during a simulation. For example, a user may want to vary a parameter in a stepwise manner to see what the operational performance trend is within the network. This is especially important during the early stages of a design when an engineer is establishing the performance bounds over an operating range of interest.

- The programs have a suite of tools for carrying out and displaying statistical analysis, signal processing, and error-rate performance evaluations. The display formats of the program can include time waveforms, electrical and optical spectra, eye diagrams, and error-rate curves.

11.3.2 FTTP Network Modeling Tool

Simulation tools are being used extensively for modeling the behavior of local, metro, and long-haul networks. These tools also can be useful in the physical layer design of FTTH/FTTP networks. The book Web site gives an application note of such a tool and related applications from RSoft Design Group. In addition, there is a link to the company Web site (www.rsoftdesign.com) that will allow readers access to an interactive demonstration model for a BPON FTTP design with 32 subscribers and a 20-km reach. Typical features of such a design example include:

- There are triple-service data/voice and video transmitters at the central office.
- The voice component is represented as a voice-over-IP (VOIP) service.
- The data/voice transmitter operates at 1.25 Gbps.
- Video is represented by a 16-QAM CATV subcarrier-multiplexed (SCM) signal.
- The total fiber length is about 20 km.
- Each of the 32 subscriber paths goes through up to three splitters and a number of splices.
- The total signal attenuation from fiber spans and splitters is about 29 dB.
- All dispersion, attenuation, and nonlinear effects in the fiber are considered.
- The simulations can be run in progressive complexity of fiber effects:
  - Fiber with loss only
  - Fiber with loss and linear effects
  - Fiber with a complete set of linear and nonlinear effects
- The triple-service ONT consists of data/VOIP and video receivers.

11.4 LINK CAPACITY ESTIMATES

A *rise-time budget analysis* is a convenient method of determining the information capacity of an optical link. This is particularly useful for a digital link in which the capacity is dispersion limited.

11.4.1 Basic Formulation

In the rise-time budget analysis approach, the total rise time t_{sys} of the link is the root-sum-square calculation of the rise times from each contributor t_i to the pulse rise-time degradation, that is,

$$t_{sys} = \left(\sum_{i=1}^{N} t_i^2\right)^{1/2} \qquad (11.5)$$

The five basic elements that may limit system speed significantly are the transmitter rise time t_{TX}, the modal dispersion rise time t_{mod} of multimode fiber, the chromatic dispersion rise time t_{CD} of the fiber, the polarization mode dispersion rise time t_{PMD} of the fiber, and the receiver rise time t_{RX}. Substituting these parameters into Eq. (11.5) then yields

$$t_{sys} = (t_{TX}^2 + t_{mod}^2 + t_{CD}^2 + t_{PMD}^2 + t_{RX}^2)^{1/2} \tag{11.6}$$

Single-mode fibers do not experience modal dispersion (i.e., $t_{mod} = 0$), so in these fibers the rise time is related only to chromatic and polarization mode dispersions.

11.4.2 Basic Rise Times

Generally, the total transition-time degradation t_{sys} of a digital link should not exceed 70 percent of an NRZ (non-return-to-zero) bit period or 35 percent for RZ (return-to-zero) data, where one bit period is defined as the reciprocal of the data rate. In Appendix D we discuss NRZ and RZ data formats in more detail.

The rise times of the transmitters and receivers generally are known to the link designer. The transmitter rise time is attributable primarily to the speed at which a light source responds to an electrical drive current. A *rule-of-thumb estimate* for the *transmitter rise time* is 2 ns for an LED and 0.1 ns for a laser diode source. The receiver rise time results from the photodetector response speed and the 3-dB electrical bandwidth B_{RX} of the receiver front end. The rise time typically is specified as the time it takes the detector output to increase from the 10 percent to the 90 percent point, as shown in Figure 11.6. If B_{RX} is given in megahertz, the *receiver front-end rise time* in nanoseconds is

$$t_{RX} = \frac{350}{B_{RX}} \tag{11.7}$$

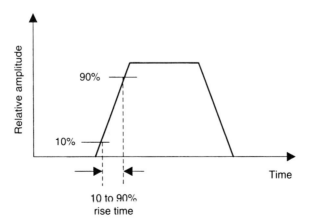

Figure 11.6. The 10 to 90 percent rise time of a pulse.

202   FTTP NETWORK DESIGN

In practice, an optical fiber link seldom consists of a uniform, continuous, jointless fiber. Instead, a transmission link nominally is formed from several concatenated (joined in tandem) fibers that may have different dispersion characteristics. This is especially true for dispersion-compensated links operating at 10 Gbps and higher. In addition, multimode fibers experience modal distributions at fiber-to-fiber joints owing to misaligned joints, different core index profiles in each fiber, and/or different degrees of mode mixing in individual fibers. Determining the fiber rise times resulting from chromatic and modal dispersion then becomes more complex than for the case of a single uniform fiber.

The fiber rise time t_{CD} resulting from chromatic dispersion over a length L can be approximated by

$$t_{CD} \approx |D_{CD}| L \Delta\lambda \qquad (11.8)$$

where $\Delta\lambda$ is the half-power spectral width of the light source and D_{CD} is the fiber chromatic dispersion. Since the chromatic dispersion value may change from one section of fiber to another in a long link, an average value should be used for D_{CD} in Eq. (11.8).

For a multimode fiber the *bandwidth*, or *information-carrying capacity*, is specified as a bandwidth–distance relationship with units of MHz·km. Thus, the bandwidth needed to support an application depends on the data rate of transmission, that is, as the data rate goes up (MHz), the distance (km) over which signals can be transmitted at that rate goes down. Multimode fibers with a 50-μm core diameter have about three times more bandwidth (500 MHz·km) than 62.5-μm fibers (160 MHz·km) at 850 nm. If B_{mod} is the modal-dispersion bandwidth (in MHz·km), the modal rise time t_{mod} (ns) over a fiber of length L (km) is given by

$$t_{mod} = \frac{440 L}{B_{mod}} \qquad (11.9)$$

The pulse-spreading t_{PMD} resulting from polarization mode dispersion is given by

$$t_{PMD} = D_{PMD} \times \sqrt{\text{fiber length}} \qquad (11.10)$$

where D_{PMD} is the polarization mode dispersion measured in units of $ps/\sqrt{km}$.

Let us now calculate the expected rise times for the link examples that are given in Section 11.2.

11.4.3 FTTP Link Rise Time

The following components are used in the FTTP link in Section 11.2:

- A laser diode with a 0.1-ns rise time and a 1-nm spectral width (this assumes that the laser has reached its stabilized *on* state, as described in Section 4.5)
- A pin-photodiode receiver with a front-end bandwidth $B_{RX} = 1250$ MHz
- A G.652 single-mode fiber with $D_{CD} = 4$ ps/(nm·km) and $D_{PMD} = 0.1$ $ps/\sqrt{km}$ at 1490 nm

TABLE 11.5 Spreadsheet for Calculating a BPON FTTP Link Rise-Time Budget

Component	Rise Time	Rise-Time Budget
Allowed rise-time budget		$T_{sys} = 0.7/B_{NRZ}$ = 0.56 ns
Laser transmitter rise time	0.1 ns	
Chromatic dispersion in fiber	80 ps	
Receiver rise time	0.28 ns	
System rise time [Eq. (11.5)]		0.10 ns

Using Eq. (11.7) yields $t_{RX} = 0.28$ ns, and Eq. (11.8) gives $t_{CD} = 0.08$ ns. Then since there is no modal dispersion in the fiber and ignoring the negligible PMD effects, the total rise time is

$$t_{sys} = (t_{TX}^2 + t_{CD}^2 + t_{RX}^2)^{1/2} = [(0.1)^2 + (0.28)^2 + (0.08)^2]^{1/2} = 0.10 \text{ ns}$$

Since the FTTP signal uses an NRZ format, the rise time needs to be less than 0.7/(1250 Mbps) = 0.56 ns. Thus, the rise-time criteria are well satisfied.

Analogous to power-budget calculations, a convenient procedure for keeping track of the various values in the rise time is to use a tabular or spreadsheet form. Table 11.5 shows an example of this for the calculation above.

11.5 NETWORK PROTECTION SCHEMES

ITU-T Recommendation G.983.5 describes the functions needed to extend ITU-T Rec. G.983.1 to enable survivability and network protection enhancements for delivering highly reliable services. The document includes PON survivability architectures, protection performance criteria, and protection-switching criteria and protocols.

As Figure 11.7 illustrates, three basic protection architectures are denoted by type A through type C. Only the optical fiber is protected with the type A scheme. The simplicity of the protection scheme shifts some of the link reconfiguration burden of operational protection to the transmission equipment. For example, the optomechanical mechanism that switches a malfunctioning fiber over to a protection fiber functions independent of any transmission protocol, such as ranging. This means that the ranging procedure needs to be done again after a type A switchover occurs.

The type B protection scheme has duplicate OLT equipment in the central office. Each OLT is connected to the PON optical splitter with an independent fiber. One OLT is designated as the primary actively working device and the other OLT serves as a hot standby. This means that if the active fiber path is broken or the primary OLT interface fails, the service is switched over to the standby OLT and link. The switchover control is done only within the central office.

The type C protection mechanism uses a fully redundant backup PON network. In this case both the primary and the backup are working. This allows for a switchover time comparable to the 50-ms maximum specified for SONET and SDH networks. The

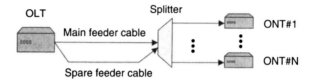

Type A (spare feeder fiber)

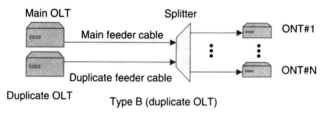

Type B (duplicate OLT)

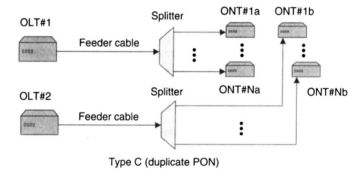

Type C (duplicate PON)

Figure 11.7. Three basic FTTP network protection architectures.

receiving equipment at either end normally selects the signal from the primary PON. However, it compares the fidelity of the signals from each PON continuously and chooses the alternative backup signal in case of severe degradation or loss of the primary signal. Thus, each ONT path is switched individually based on the quality of the signal received. As a result, the type C protection method can protect against fiber failures in the feeder, distribution, or drop cables and against OLT or ONT failures.

11.6 SUMMARY

The design of high-quality links for FTTP optical distribution networks involves a series of trade-offs among the many interrelated performance variables of each component based on the system operating requirements.

In carrying out a link power-budget analysis, one first needs to determine the link power margin or allowed attenuation range. Simply, this is the maximum allowed difference in light power level between the optical transmitter output and the minimum receiver sensitivity needed to establish a specified BER. This margin can be allocated to fiber, splitter, splice, and connector losses, plus any additional system margin, which is an optical-power safety factor for link design. The system margin decision involves adding extra decibels to the power requirements to compensate for possible unforeseen link degradations. These degradations could arise from factors such as a dimming of the light source over time, aging of other components in the link, the possibility that certain splices or connectors in the actual link have a higher loss than anticipated, or additional losses occurring when a cable is repaired.

Certain operational factors in a link usually contribute to signal impairment. Among these factors are modal noise, chromatic dispersion, polarization-mode dispersion, reflection noise in the link, low extinction ratios in the laser, or frequency chirping. When any of these dispersion or nonlinear effects contribute to signal impairment, there is a reduction in the signal-to-noise ratio (SNR) of the system from the ideal case. This reduction in SNR is known as the power penalty for that effect, which generally is expressed in decibels.

ITU-T Recommendation G.983.5 describes the functions needed to extend ITU-T Rec. G.983.1 to enable survivability and network protection enhancements for delivering highly reliable services. The document includes PON survivability architectures, protection performance criteria, and protection-switching criteria and protocols.

PROBLEMS

11.1 Suppose that a G.655 optical fiber used in an FTTP network has a chromatic dispersion of $D_{CD} = 4.0 \times 10^{-3}$ ns/(nm·km) at 1490 nm. What is the power penalty due to chromatic dispersion over a 20-km distance for a 2.5-Gbps data rate if the spectral width of the optical source is 0.1 nm?

11.2 Plot the power penalties in an FTTP link for extinction ratios ranging from 5 to 20. What is the power penalty when the extinction ratio is 10?

11.3 Consider the 1310-nm power-budget design example given in Section 11.2.2. What is the operating margin if the following conditions change: a 1×16 optical power splitter that has a throughput loss of 13.5 dB is used and now there are five optical connectors and four splices, each of which has a 0.1-dB loss? All other parameters are the same.

11.4 Using the conditions listed in Problem 11.3, what is the power-budget operating margin for the 1490-nm link described in Section 11.2.3?

11.5 The book Web site (see page xvi) gives a link to an example of a commercially available simulation tool for carrying out link designs of FTTP networks. Check out the characteristics of this design tool and run through some examples of FTTP link performance simulations.

206 FTTP NETWORK DESIGN

11.6 The 1310-nm power-budget design example given in Section 11.2.2 and Table 11.3 uses very conservative loss numbers for the link components. Suppose that the design needs to have a 3.0-dB safety margin. In addition, assume that in practice there may be an additional loss of 1.8 dB from other connectors and splices and let the extinction ratio power penalty be 0.8 dB. What is the realizable transmission distance at 1310 nm for a 1 × 32 optical splitter architecture when these additional losses are included in the power budget?

11.7 For the power-budget design example given in Section 11.2.2, what effect do variations in component losses have on the achievable transmission distance? These loss variations could be due to factors such as dirty or misaligned connectors, additional unforeseen splices, tight fiber bends in equipment cabinets, or degradations in the transmitter or the receiver.

11.8 What are the optical losses between point A and points B, C, D, and E in Figure 11.8? Assume that there are six connector or splice joints with a loss of 0.1 dB each, the fiber has a 0.4-dB/km attenuation, and the 1 × 8 and the 1 × 4 optical power splitters have throughput losses of 10.7 and 7.7 dB, respectively.

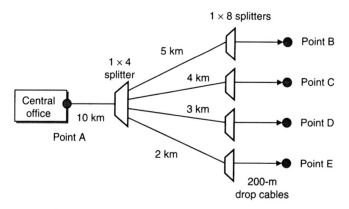

Figure 11.8. FTTP network layout for Problem 11.8.

11.9 Draw the details of the internal optical path in an optical power splitter for the type B network protection scheme shown in Figure 11.7.

11.10 Describe the protection-switching process that might be implemented in a type C protection method (see Figure 11.7) under the following failure conditions:
(a) ONT#2a fails.
(b) ONT#2a and ONT#15b fail at approximately the same time.

FURTHER READING

1. A. B. Carlson, *Communication Systems*, 4th ed., McGraw-Hill, Burr Ridge, IL, 2002.
2. O. Gerstel and R. Ramaswami, "Optical layer survivability: a services perspective," *IEEE Commun. Mag.*, vol. 38, pp. 104–113, Mar. 2000.
3. S. Haykin, *Communication Systems*, 4th ed., Wiley, New York, 2000.
4. ITU-T Recommendation G.983.5, "A Broadband Optical Access System with Enhanced Survivability," Jan. 2002.
5. G. Keiser, *Optical Communications Essentials*, McGraw-Hill, New York, 2003.
6. M. D. Vaughn, D. Kozischek, A. Boskovic, and R. E. Wagner, "Value of reach-and-split ratio increase in FTTH access networks," *J. Lightwave Technol.*, vol. 22, pp. 2617–2622, Nov. 2004.

12

FTTP NETWORK IMPLEMENTATIONS

Designing and deploying FTTP networks requires careful evaluation of all communication links between the transmission equipment in the *central office* (CO) and the ONT at the customer's premises. Network layout considerations within the CO include where to place the WDM coupler that combines the triple-play services, minimizing the number of connector interfaces, and allowing flexibility in connecting both FTTP and non-FTTP services to available fibers in the feeder cables. Installing the *outside cable plant* (OSP) is a major expense of an FTTP implementation. After the optical cable is in place, it is expensive and difficult to replace or retrofit it. In addition, choosing the appropriate number of fibers in the cable is important for allowing the possibility of other services to use spare fibers and for future upgrades or expansions.

In this chapter we first present typical cabling interfaces and the associated splicing and connector enclosures in the central office. We then give examples of what types of optical fiber cables are used to connect different parts of the OSP. In addition, we describe the associated splice boxes and cable distribution cabinets located in the OSP. These need to be environmentally rugged enclosures and cabinets for housing the optical power splitters and the splice holders that serve as interfaces between the feeder, distribution, and drop cables. Since the splitter divides the optical power for delivery to individual premises, a multiple-fiber cable leaves the splitter distribution cabinet and terminates in a splice enclosure. This enclosure can be either an aerial installation on a pole, or it can be located in an aboveground or underground cabling vault. From there, individual drop cables run to the customer sites.

12.1 CENTRAL OFFICE CONFIGURATION

Key factors concerning the cable layout in the CO include selecting the connection routes between the transmission equipment and the termination point of the feeder

---

FTTX Concepts and Applications, by Gerd Keiser
Copyright © 2006 John Wiley & Sons, Inc.

cable, providing access points for performance testing, and determining the optimum location for the WDM coupler used with triple-play voice, video, and data equipment.

12.1.1 Service Inputs to the FTTP Network

Figure 12.1 shows a simplified diagram of the transmission equipment in a central office that may be connected to an optical cable network. An OLT interfaces with long-haul and metro telecommunication network equipment such as telephone switches, ATM switches, various OC-N or STM-N transmission systems, Ethernet modules, and digital video equipment. The OLT combines the incoming traffic into one continuous digital stream, superimposes it onto a 1490-nm wavelength, and then sends it to an optical multiplexer. Transmission of analog video traffic over a PON is carried out by superimposing the analog video signal onto a 1550-nm wavelength and then combining it in a wavelength multiplexer onto the same downstream fiber as the 1490-nm voice and data stream.

Depending on the manufacturer and the OLT size and capability (e.g., how many PONs it can serve), an OLT can be configured for a wide variety of applications through the use of different plug-in modules. The OLT typically is housed in some type of 2-m (7-ft)-high equipment rack. Although the analog video equipment could be in the same equipment rack as the OLT, often it is located in a different nearby rack.

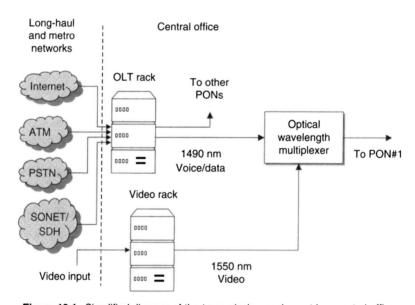

Figure 12.1. Simplified diagram of the transmission equipment in a central office.

12.1.2 Cable Layout and Interfaces

Figure 12.2 shows a simplified cabling network in a central office. The main elements include racks containing OLTs or video equipment, fiber distribution frames, fiber entrance cabinets, and various types of indoor optical fiber cables. A *fiber distribution frame* (FDF) is a rack-sized structure that enables a modular and flexible cable interconnection network within a central office. Basically, the FDF is a large panel of interconnected fiber termination points that can be hooked together with short patch cords. This large panel of connectors allows easy changes in the configuration of a CO cabling network as the CO expands or when telecommunication services are added or dropped. Modern central offices tend to use a cross-connect FDF system, as Figure 12.3 illustrates. In such a configuration, both the optical fibers entering the CO from the outside and the fibers from the central office–based FTTP equipment are attached to the back of the FDF. A service technician then uses short *cross-connect patch cords* on the front panel of the FDF to interface any piece of equipment to an OSP fiber.

In practice, there is a *fiber entrance cabinet* where the fibers from the OSP are terminated. Figure 12.4 shows an entrance cabinet, which could be mounted on a

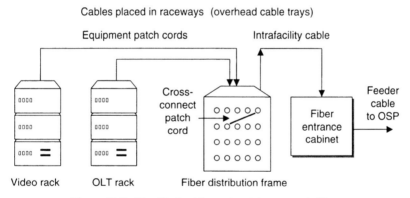

Figure 12.2. Simplified cabling network in a central office.

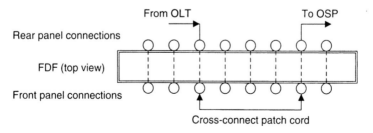

Figure 12.3. Concept of implementing a cross-connect FDF system.

212   FTTP NETWORK IMPLEMENTATIONS

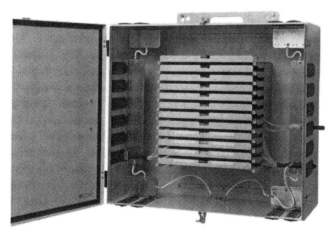

Figure 12.4. Wall-mounted entrance cabinet. (Photo courtesy of ADC; www.adc.com.)

wall or placed in an equipment rack. Since an incoming OSP cable could contain hundreds of fibers, the entrance cabinet must have a well-designed management scheme for easy splicing, identification, and storing of OSP fibers. The cabinet should also include splice trays that feature bend-radius protection of the fibers.

Now let us see how all the cable network elements are interconnected. Starting at the transmission racks in Figure 12.2, *equipment patch cords* attach the OLT and the video overlay equipment to the back side of the FDF. Coming from the other direction, *intrafacility fibers* are spliced to the OSP fibers at the entrance-cabinet point and run to the back of the fiber distribution frame. At the FDF, both the equipment patch cords and the intrafacility fibers are terminated in optical connector receptacles.

By using the setup shown in Figure 12.2, the central office FTTP transmission equipment can be connected to the appropriate OSP cable fibers by means of short optical fiber *cross-connect patch cords* on the front side of the FDF. These patch cords have optical fiber connectors on both ends to allow for easy setup and reconfiguration of the FTTP equipment. Some details about various optical fiber patch cords and intrafacility cables used in a central office are given in Section 12.1.4.

12.1.3   WDM Coupler Placement

The function of the WDM coupler for triple-play FTTP services is to combine and separate the various 1310-, 1490-, and 1550-nm signals onto and off a single optical fiber. Since video transmission in FTTP networks is handled by a separate overlay wavelength, the voice/data OLT and the video transmitter normally are in separately located equipment racks within a central office. The question then arises as to where to place the WDM coupler. Two possible locations are in the OLT equipment rack or in the FDF.

Figure 12.5 shows the cable connections in the CO if the WDM coupler is placed in the OLT equipment rack. In this case equipment patch cables are used to attach the OLT and the video transmitter to the input connectors of the coupler. An

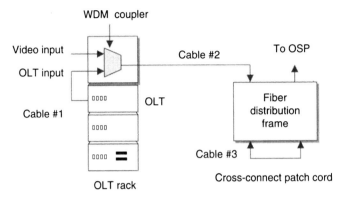

Figure 12.5. Cable connections in the CO if the WDM coupler is in the OLT equipment rack.

intrafacility cable then connects the WDM coupler to the back of the fiber distribution frame. A final cross-connect patch cord is used on the front of the FDF to complete the coupler output path to the OSP feeder fiber. Thus, a total of three fiber cable segments are used when the coupler is located in the OLT rack.

Figure 12.6 illustrates the situation when the WDM coupler resides in the back panel of the FDF. Now intrafacility cables are used to attach the OLT and the video transmitter to the input connectors of the coupler. Again a final fiber patch cord is used on the front of the FDF to complete the coupler output path to the OSP feeder fiber. Thus, a total of two fiber cable segments are used in either path when the coupler is located in the OLT rack. Since the WDM coupler already resides in the FDF, there is no need for the coupler-to-FDF cable used in Figure 12.5.

In addition to requiring an extra cable when the WDM coupler is in the OLT rack, and thus having an extra connector loss in the link, this configuration also reduces the central office cabling flexibility. For example, if the OLT rack needs to be moved, modified, or replaced, all the video connections to the WDM couplers also need to be changed. Thus, it is more advantageous to place the WDM coupler in the back side of the FDF.

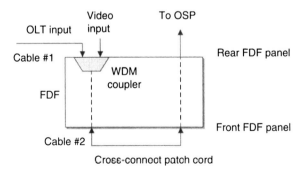

Figure 12.6. Cable connections in the CO if the WDM coupler resides in the back panel of the FDF.

TABLE 12.1 Cable Markings Based on the U.S. National Electrical Code and UL Test Specifications

Marking	Cable Name	UL Test	Substitute
OFNP	Optical fiber nonconductive plenum	UL 910	None
OFCP	Optical fiber conductive plenum	UL 910	None
OFNR	Optical fiber nonconductive riser	UL 1666	OFNP
OFCR	Optical fiber conductive riser	UL 1666	OFCP
OFN	Optical fiber nonconductive	UL 1581	OFNP or OFNR
OFC	Optical fiber conductive	UL 1581	OFCP or OFCR

12.1.4 Patch Cords and Intrafacility Cables

Three types of fiber cables are commonly used within a central office. These are intrafacility cables, equipment patch cords, and cross-connect patch cords. An important factor for using a cable in a building is the *flammability rating*. The National Electrical Code (NEC) in the United States establishes flame ratings for cables, while on a global scale, Underwriters' Laboratories (UL) has developed cable test procedures. For example, depending on the application, the NEC requires that all cables which run through a specific region must be constructed of a certain grade of low-smoke and fire-retardant materials. In addition, since the intrafacility cables and patch cords are used within a building, the NEC rules state that the cables must be marked correctly and be installed properly in accordance with their intended use.

Table 12.1 lists some common indoor cable applications, their markings, and the UL flammability test requirements. As NEC Article 770 describes, the outer protective cable jacket material will vary based on the particular application. The three basic indoor building regions in which cables may be placed are known as the plenum, riser, and general-purpose areas.

- *Plenum cables.* A *plenum* is the empty space within walls, under floors, or above drop ceilings used for airflow, or it can form part of an air distribution system used for heating or air conditioning. Plenum-rated cables are UL-certified by the UL-910 *plenum fire test method* as having adequate fire resistance and low smoke-producing characteristics for installations in these spaces without the use of a conduit. As noted in Table 12.1, these cables are termed OFNP (*optical fiber nonconductive plenum*) for all-dielectric cables or OFCP (*optical fiber conductive plenum*) when they contain metallic components.

- *Riser cables.* A *riser* is an opening, shaft, or duct that runs vertically between one or more floors. Riser cables can be used in these vertical passages. Riser-rated cables are UL-certified by the UL-1666 *riser fire test method* as having adequate fire resistance for installation without conduit in areas such as elevator shafts and wiring closets. As noted in Table 12.1, these cables are termed OFNR (*optical fiber nonconductive riser*) or OFCR (*optical fiber conductive riser*). Note that plenum cables may be substituted for riser cables, but not vice versa.

- *General-purpose cables.* A general-purpose area refers to all other regions on the same floor that are not plenum or riser spaces. General-purpose cables can be installed in horizontal, single-floor connections, for example, to connect from a wall jack to a computer. However, they cannot be used in riser or plenum applications without being placed in fireproof conduits. To qualify as a general-purpose cable, it must pass the UL 1581 *vertical-tray fire test*. As shown in Table 12.1, these cables are rated OFN (*optical fiber nonconductive*) or OFC (*optical fiber conductive*). Note that plenum or riser cables may be substituted for general-purpose cables, but not vice versa.

The number of fibers in an *intrafacility cable* (IFC) can range from 12 to 216. A major application is for the transmission path between the fiber entrance cabinet and the FDF. Standard intrafacility cables are rated for riser applications and meet the UL 1666 OFNR flammability requirements. The fiber packaging within an IFC can be of either a ribbon cable format or can consist of strands of tight-buffered fibers (see Section 5.3). In either case, all fibers are color-coded for easy identification.

The two types of patch cords are similar in nature but may vary in their length and connector styles. *Equipment patch cords* are used for the interface between the FTTP OLT or video transmission equipment and the back of the FDF. A nominal configuration is a two-fiber cable, popularly known as a *zip cable*, with a length of 10 m. It can have optical connectors on both or just one of the ends. For the latter case, the end with the connectors is used at the equipment interface and the other end can be spliced onto the rear of the FDF to provide a lower-loss interface than a connector can offer.

A cross-connect patch cord is known popularly as a *jumper cable*. It has a connector on both ends and generally is about 2 m long. Its main application is to provide the front-panel interconnections on the FDF.

12.2 FEEDER CABLES

A fundamental design consideration for FTTP networks is the specification of an optical fiber cable that is appropriate for the neighborhood in which it is to be used. This includes not only selecting a cable structure that is suitable for the specific environment in which it will be installed, but also determining how many fibers should be contained in the cable. The fiber count will allow the network owner to establish multiple FTTP networks for current use and for future expansion to more PONs. In addition, spare fibers in the feeder cables may be used for other applications which are not related to the FTTP network.

12.2.1 Feeder Cable Structures

An individual FTTP network normally is designed to serve up to 32 homes or businesses. At least one feeder fiber is needed between the central office and the optical splitter in order to reach these 32 premises. Thus, to serve a community that has a large number of customer sites, say 1600 premises, a feeder cable will contain at least 1600/32 = 50 fibers. In addition, there usually are extra fibers in the feeder cable for redundancy and future expansion purposes. A variety of other factors can increase the fiber count. For example, since a feeder cable usually passes businesses and organizations that are not part of an FTTP network, spare fibers in the feeder cable can be used to carry services from other networks to these customers. This means that

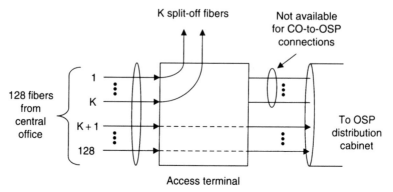

Figure 12.7. Path configuration of a 128-fiber feeder cable running from a central office to an outside distribution cabinet.

within the FDF in the central office, each fiber in the feeder cable needs to be easily identifiable and accessible to offer the greatest amount of cable usage flexibility.

Figure 12.7 shows a generic scenario for this. Here a 128-fiber feeder cable runs from a central office to an outside distribution cabinet. Along the way, a group of K fibers is split off in a fiber access terminal to serve other networks. These could be PON clusters, high-capacity users requiring point-to-point gigabit Ethernet links, or large file storage farms. Note that once this group of K fibers is split off from the cable, their continuation segments no longer can be used as a path between the CO and the far end OSP distribution point. Although in theory a cable with a different fiber count could be used starting at the split-off point, in an actual network it may be easier to use the same type of cable from a logistics point of view. However, the continuation fibers could be used for transmission paths between the split-off point and nodes farther along the path. An alternative to having a large feeder line is to use cables with smaller fiber counts and run them through microducts that are placed within the larger duct (see Section 12.4).

12.2.2 OSP Distribution Cabinet

The function of the OSP *distribution cabinet* is to be the termination point of the feeder cables in the OSP, to house the optical splitters and splice boxes, to serve as a convergence point for the distribution cables running to clusters of premises, and to protect these components from the environment. Standard features of the OSP distribution cabinet include:

- Resistance to moisture, wind, rodents, intruders, and fire
- Easy access but with wind-resistant doors
- A neutral shape and color so that it is not an eyesore
- Adherence to Telcordia requirements for outside equipment (see Appendix E)

Figure 12.8 shows an example of an OSP distribution cabinet, which has the capacity to hold at least 100 optical splitters (serving 100 separate FTTP networks). This means that the cabinet must have an access panel capable of terminating at least

Figure 12.8. OSP distribution cabinet. (Photo courtesy of ADC; www.adc.com.)

100 fibers coming from a feeder cable. If a splitter serves up to 32 customers, a total of 3200 premises can be connected to these 100 splitters.

12.3 DISTRIBUTION SECTION

Fibers leaving the OSP distribution cabinet are destined for individual premises. A wide range of distribution schemes are possible, depending on the layout of the neighborhood being served. The main elements in the distribution section are distribution cables, access terminals or enclosures, and drop cables.

Figure 12.9 shows a cabling layout in which four clusters of eight homes are served by an FTTP network using a 1×32 optical splitter. Here a 32-fiber cable leaves the OSP distribution cabinet and is mounted on poles or runs underground to connect up to 32 premises. There are four breakout points along the cable path where groups of eight fibers are split off for connecting to eight premises.

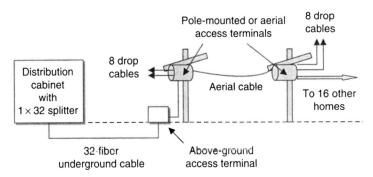

Figure 12.9. Distribution cable layout to four clusters of eight homes.

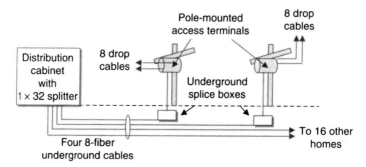

Figure 12.10. Four eight-fiber distribution cables servicing four clusters of eight homes.

At a breakout point there is an enclosure called an *access terminal*, which is the interface between the distribution and the drop cable segments. These enclosures can be located either underground, on poles, in aboveground pedestal shelters, or in an indoor wiring closet in a large apartment or office building. The enclosures are available in many sizes, with different physical characteristics depending on the environment where they are located. The *drop cables* complete the path between the access terminal and the customer premises. An alternative design is to have four eight-fiber distribution cables run from the distribution cabinet to an environmentally rugged optical fiber cross-connect box near each cluster of eight homes, as Figure 12.10 illustrates.

Figure 12.11 shows a cross-connect enclosure that can be used for aerial, pole-mounted, or manhole applications. Such enclosures have built-in holders for

Figure 12.11. Environmentally rugged optical fiber cross-connect enclosure for aerial, pole-mounted, or manhole applications. (Photo courtesy of ADC; www.adc.com.)

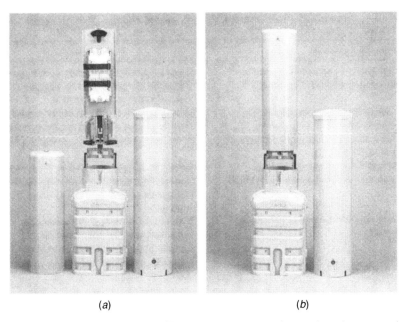

Figure 12.12. Aboveground optical fiber cross-connect enclosure based on a pedestal design: (*a*) inner dome off; (*b*) inner dome on. (Photo courtesy of Charles Industries; www.charlesindustries.com.)

storage of slack fiber lengths and typically can accommodate up to 96 splices. Within the enclosure connections can be made between feeder and distribution fibers or between distribution and drop fibers using short (nominally 1.5 m) patch cords.

Figure 12.12 shows an aboveground cross-connect enclosure based on a pedestal design. In a fielded installation the square section at the bottom of the enclosure is placed in the ground to anchor the unit in place. The inner dome shown in the figure creates a watertight seal that keeps out moisture, dirt, and debris. The larger exterior dome creates a flood-protection barrier and an additional layer of environmental protection. Such units come in a variety of sizes and can accommodate from 24 to 288 splices.

12.4 INSTALLATION OF PON CABLES

A great number of PON installations take place in urban or suburban neighborhoods, since the span of such a network is typically less than 20 km. In such environments cables are installed by a method suitable for that situation. These include pulling or blowing the cable through underground ducts, burying it in an outside trench, plowing the cable directly into the ground, suspending it on poles, or drilling an underground path for the cable to pass through. Although each method has its own special handing procedures, they all need to adhere to a common set of precautions. These include avoiding sharp bends in the path, minimizing stresses on the installed cable,

TABLE 12.2 Some ITU-T Recommendations for Optical Fiber Cable Installation Techniques

Recommendation Name	Description
L.35: "Installation of optical fiber cables in the access network," Oct. 1998.	Gives guidance for installing optical cables in ducts, on poles, and through direct burial in the access network.
L.38, "Use of trenchless techniques for the construction of underground infrastructures for telecommunication cable installation," Sept. 1999.	Describes drilling techniques for installing underground telecommunication cables without the need for excavation or plowing.
L.39, "Investigation of the soil before using trenchless techniques," May 2000.	Describes soil-investigation techniques for obtaining information about the position of buried objects and the nature of the ground.
L.42, "Mini-trench installation techniques," Mar. 2003.	Describes a technique that allows the installation in small trenches of underground optical cables in ducts. The advantages are speed of execution, lower cost, significantly lower environmental impact, and limited surface disruption
L.43, "Micro-trench installation techniques," Mar. 2003.	Describes installing underground cables at a shallow depth in small grooves.
L.57, "Air-assisted installation of optical fiber cables," May 2003.	Describes air-assisted methods for installation of optical fiber cables in ducts.

periodically allowing extra cable slack along the route for unexpected repairs, and avoiding excessive pulling forces or sudden hard tugs on the cable.

The ITU-T has published a number of L-series recommendations for the construction, installation, and protection of fiber optic cables in the outside plant. Table 12.2 lists some of these and states their application.

12.4.1 Direct-Burial Installations

In *direct-burial installation* methods a fiber optic cable is either plowed directly underground or buried in an outside trench. Different burial depths are required for a fiber optic cable, depending on where in the FTTP network it is to be installed. Table 12.3 lists some depth requirements given in the TIA/EIA-590-A standard.

TABLE 12.3 Installation Depth Requirements for Fiber Cable Given in TIA/EIA-590-A

FTTP Network Segment	Minimum Installation Depth [in. (cm)]
Feeder or distribution cable	24 (60)
Drop cable	18 (45)
Underground ducts	30 (75)

(a)

(b)

Figure 12.13. Two scenes of a plowing operation that may be carried out in a suburban area: (a) an easy open-space operation; (b) a tight squeeze between trees. (Photos courtesy of Vermeer Manufacturing Company; www.vermeer.com.)

Figure 12.13 illustrates two scenes of a *plowing operation* that may be carried out for installing drop cables in a suburban area. The cables are mounted on small reels on the plowing vehicle and are fed directly into the ground by means of the plow mechanism. The vehicle shown here is a compact unit that is 35.5 in. (90 cm) wide. This size allows it to fit through tight openings such backyard gates or narrow spaces between bushes or trees in a lawn area, as illustrated in the bottom scene. Such a machine can plow a cable into the ground to depths of 18 in. (46 cm). Smaller and significantly larger Caterpillar-mounted plows also are available for feeder and distribution cable installations.

In cases where a plowing operation may not be feasible or desirable because of terrain conditions or a particular soil characteristic, a *trenching* method can be used. This method is more time consuming than direct plowing since it requires placing and burying the cable in a trench that first has to be dug to some specified depth. However, trenching allows the installation to be more controlled than in plowing. For example, in direct plowing it is not known if a sharp rock is left pressing against the installed cable or if the cable was damaged in a way that may cause it to fail later. In a suburban area this digging can be done with a machine such as the one shown in Figure 12.13, where the plow is replaced with a trenching mechanism. Figure 12.14 shows a tractor-mounted trenching attachment for urban or suburban use. Such a machine can dig a 5- to 12-in. (13- to 30-cm) wide trench to depths of 60 in. (152 cm). The tractor itself offers a wide application flexibility, since the trenching attachment can be replaced by either a plow or a backhoe. In addition, as

Figure 12.14. Tractor-mounted trenching attachment for urban or suburban use. (Photo courtesy of Vermeer Manufacturing Company; www.vermeer.com.)

can be seen in the picture, the blade on the front of the tractor serves as an earth-moving unit.

12.4.2 Horizontal Drilling

A technique called *directional boring* or *horizontal drilling* may be needed in areas where the surface cannot be disturbed. For example, if the cable path needs to cross a busy road, it is better to drill an underground duct-encapsulated hole under the road and then run the cable through that duct. This technique is in wide use for installations such as water mains, electric lines, gas pipes, and telecommunication cables. The drilling machines come in at least a dozen different sizes, depending on the depth and distance that holes need to be bored. For example, the horizontal drilling machine illustrated in Figure 12.15 is a compact unit that is 35.5 in. (90 cm) wide for suburban use. This particular machine can bore a hole for

Figure 12.15. Compact horizontal drilling machine for suburban use. (Photo courtesy of Vermeer Manufacturing Company; www.vermeer.com.)

Figure 12.16. Compact horizontal drilling machine for urban use. (Photo courtesy of Vermeer Manufacturing Company; www.vermeer.com.)

a 1.66-in. (4-cm)-diameter pipe below the surface for distances of up to 198 ft (60.4 m).

Figure 12.16 shows a mid-sized drilling machine for urban use, such as drilling underneath streets or creeks. This 48-in.-wide machine can bore a hole for a 1.9-in. (5-cm)-diameter pipe below the surface for distances of up to 400 ft (122 m).

12.4.3 Pulling Cable into Ducts

Most ducts used for telecommunication cables are constructed of a high-density polyurethane, PVC, or an epoxy fiberglass compound. A traditional installation procedure is to pull a cable through such a duct by means of a special rope attached to one end of the cable. To reduce pulling tensions during cable installation, the inside walls can have longitudinal or corrugated ribs or they may have been lubricated at the factory. Alternatively, a variety of lubricants are available that may be applied to the cable itself as it is pulled into a long duct or one that has numerous bends.

Ducts also can contain a factory-installed pulling tape running along its length. This is a flat tape similar to a measuring tape that has markings every meter for easy identification of distance. If the duct does not contain a pulling tape, it can be fished through or blown into a duct length. After the fiber optic cable is installed in a duct, end plugs can be added to prevent water and debris from entering the duct.

12.4.4 Cable Jetting Installation

An alternative method to a pulling procedure is to employ a high-pressure airflow to blow a fiber cable into a duct. The installation scheme of utilizing the friction of forced air pushing on the cable jacket is referred to as an *air-assisted* or a *cable jetting* method. Cable jetting must overcome the same frictional forces to move a cable

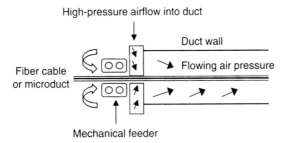

Figure 12.17. Mechanical rollers and air pressure are the driving forces in the cable jetting method.

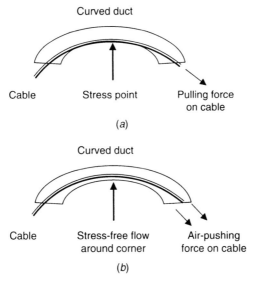

Figure 12.18. (a) The pulling method puts a high lateral stress on the cable at duct bends. (b) With cable jetting the cable moves freely around bends in a duct.

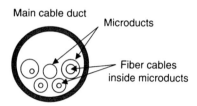

Figure 12.19. The concept of using multiple miniature microducts within a larger conduit.

TABLE 12.4 Maximum Number of Microducts That Can Fit into a Particular Main Duct

Main Duct Size	7	10	12
	Microduct Outer Diameter (mm)		
25	3	1	0
40	10	5	4
50	20	10	8

as in a pulling operation, but it does this differently and with much less mechanical stress on the cable.

As Figure 12.17 shows, the two driving forces in the cable jetting method come from a mechanical roller mechanism that pushes the cable into the duct and from the force of air pushing on the cable jacket. The advantage of cable jetting is that the cable moves freely around bends, whereas the pulling method puts a high lateral stress on the cable when it is passes through bends in a duct. This is illustrated in Figure 12.18. The top part of the figure shows that during a pulling operation a high lateral stress is imposed on a cable at bends in the duct. With a cable jetting procedure such lateral forces are mitigated since the forced air pushes on the cable from all sides. This tends to keep the cable in the middle of the duct as it is blown through, as the bottom part of Figure 12.18 shows.

The use of multiple miniature tubes or polyethylene microducts placed within a larger conduit allows an incremental installation of fiber cables in a duct over a period of time. Figure 12.19 illustrates this concept. The main duct inner diameter can range from 0.75 to 2 in. (27 to 60 mm), and standard microduct outer diameters are 7, 10, and 12 mm. A cable jetting method can be used to install the microducts within new or existing ducts and also for blowing the fiber into the microducts. Table 12.4 lists the maximum number of 7-, 10-, and 12-mm microducts that can be blown into different sizes of main ducts; for example, five 10-mm ducts fit into a 40-mm main duct.

Figure 12.20 shows an example of a cable jetting machine. The cable or duct is fed in from the top right. The central part is the mechanical feeder and the air-pressure connection pipe is seen in the leftmost section. Cables or microducts can be installed at rates of 150 to 300 ft per minute with up to 20 corners in a duct. A smaller cable

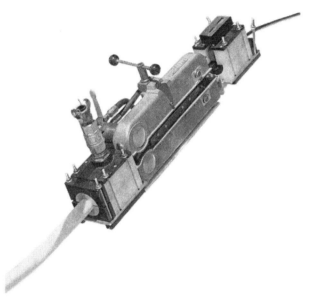

Figure 12.20. Cable jetting machine. (Photo courtesy of Sherman & Reilly, Inc; www.sherman-reilly.com.)

Figure 12.21. A small cable jetting machine can be used for FTTP applications. (Photo courtesy of Sherman & Reilly, Inc; www.sherman-reilly.com.)

jetting machine such as the one shown in Figure 12.21 can be used to install cables within a microduct. This is particularly useful for FTTP applications, such as installations of drop cables to customer premises. The distance that a fiber cable or microduct can be blown depends on their size, the inner diameter of the main duct, the air pressure generated by the compressor, and how many bends or turns there are

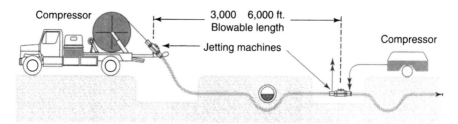

Figure 12.22. Jetting machines can be cascaded for long installation lengths. (Diagram courtesy of Sherman & Reilly, Inc; www.sherman-reilly.com.)

in the main duct. Thus, the distances for a single-stage cable jetting operation can range from 1 to 1.8 km.

For longer installation lengths the jetting machines can be cascaded, as Figure 12.22 illustrates. This figure shows how a cable is fed from a truck-mounted reel into a cable jetting machine at the beginning of a large duct. After a distance of 1 to 1.8 km (about 3000 to 6000 ft), a second cable jetting machine and its associated air compressor and other peripheral equipment continue the installation.

12.4.5 Aerial Installation

Cable crews can install an aerial cable either by lashing it onto an existing steel *messenger wire* that runs from pole to pole or by suspending the cable directly between poles if it is a self-supporting design. Several different methods can be used to install the fiber optic cables. The primary method for installing self-supporting cable is a *stationary reel technique*. This method stations the payoff reel at one end of the cable route and the take-up reel at the other end. A pull rope is attached to the cable and is threaded through pulleys on each pole. The take-up reel gradually pulls the cable from the payoff reel, the pulleys guide it into position along the route, and it is then attached to the poles.

If a *messenger wire* is used, this wire is first installed between poles with an appropriate tension and sag calculated to support the fiber optic cable. The messenger wire must be grounded properly and should be kept on one pole side along the route whenever possible. One of at least three techniques can then be used to attach the fiber optic cable to the messenger wire. Each of these methods uses a special lashing machine that hangs on the messenger wire and attaches the cable to the messenger as it moves along the wire length.

12.4.6 Cable Warning and Identification Markers

During direct-burial installations a standard procedure is to place a *warning tape* in the ground about 18 in. (about 50 cm) below the surface for feeder and distribution cables to alert future digging operators to their presence. The standard color code of this tape for telecommunication cables, such as fiber optic cables, is bright orange. The tape may contain metallic strips so that it can be located from aboveground with

INSTALLATION OF PON CABLES   229

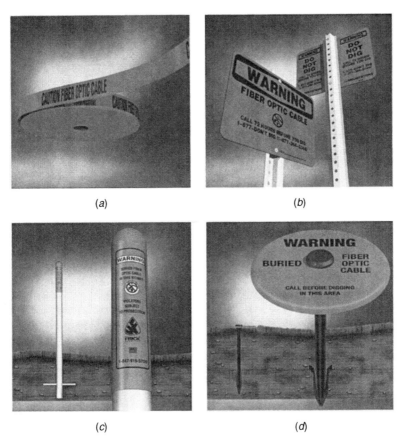

Figure 12.23. (*a*) Underground tapes; (*b*) metal posts; (*c*) flexible posts; and (*d*) ground-flush markers are used to indicate that a cable is buried in that location. (Photos courtesy of William Frick & Company; www.fricknet.com.)

a metal detector. Figure 12.23 shows an example of a warning tape and the words printed on it.

In addition, Figure 12.23 illustrates some other permanent cable location and identification methods. A cable marker that is flush with the ground may be used to indicate where a cable is buried. Other options are a visible warning post made of durable plastic or an aluminum sign with a baked enamel finish that is bolted to a steel post. All these markers can contain text and graphic information such as the cable route, cable composition, company name and logo, and emergency contact phone numbers. A standard color format for telecommunication cables is black lettering on a bright orange background. Besides indicating to repair crews where a cable is located, these precautions also are intended to minimize the occurrence of what is known popularly in the telecommunications world as *backhoe fade*. This refers to the rupture of a cable by a backhoe operator who may be unaware that a cable is located in the digging area.

230 FTTP NETWORK IMPLEMENTATIONS

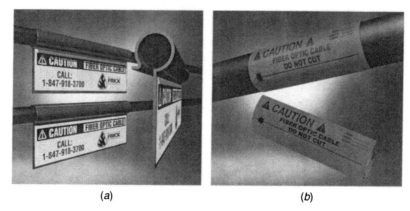

(a)   (b)

Figure 12.24. (*a*) Overhead and (*b*) underground cable markers. (Photos courtesy of William Frick & Company; www.fricknet.com.)

A variety of durable labels are available to identify what a particular underground or overhead cable is and what its characteristics are. Figure 12.24 shows a brightly colored preprinted polyethylene marker for underground, corrosive, and moist environments. Figure 12.24 also shows a marker intended for overhead cable identification. Such a marker may be made from baked enamel aluminum or stainless steel. These labels come in different sizes and can contain text and graphic information such as the type of cable, company name and logo, and emergency contact phone numbers.

12.5 SUMMARY

Designing and deploying FTTP networks requires careful evaluation of all communication links between the transmission equipment in the central office (CO) and the equipment at the customer premises. Network layout considerations within the CO include WDM coupler location, minimizing the number of connector interfaces, and allowing flexibility in connecting both FTTP and non-FTTP services to available fibers in the feeder cables. Installing the outside cable plant (OSP) is a major expense of an FTTP implementation. After the optical cable is in place, it is expensive and difficult to replace or retrofit it. In addition, choosing the appropriate number of fibers in the cable is important for allowing the possibility of other services to use spare fibers and for future upgrades or expansions.

The main network elements in a central office include transmission equipment racks, fiber distribution frames, fiber entrance cabinets, and various types of indoor optical fiber cables. A fiber distribution frame is a large panel of interconnected fiber termination points for fibers coming from the CO transmission equipment and from the OSP. These points can be hooked together with short patch cords to create complete paths between the CO equipment and the optical fibers in the feeder cable. This panel of connectors also allows easy changes in the configuration of the internal cabling network as the CO expands or when telecommunication services are added or dropped.

Fibers leaving the OSP distribution cabinet are destined for individual premises. A wide range of distribution schemes are possible, depending on the layout of the neighborhood being served. The main elements in the distribution section are distribution cables, access terminals or enclosures, and drop cables.

A great number of FTTP installations are in urban or suburban neighborhoods. In such environments cables are installed by a method suitable for that situation. These include pulling or blowing the cable through underground ducts, burying it in an outside trench, plowing the cable directly into the ground, suspending it on poles, or drilling an underground path for the cable to pass through. Although each method has its own special handing procedures, they all need to adhere to a common set of precautions. These include avoiding sharp bends in the path, minimizing stresses on the installed cable, periodically allowing extra cable slack along the route for unexpected repairs, and avoiding excessive pulling forces or sudden hard tugs on the cable.

PROBLEMS

12.1 Optical fiber cross-connect and equipment *patch cords* are essential components in a central office. These short cables are typically 2 and 10 m long, respectively. Using vendor data sheets, list some characteristics of various types of optical fiber patch cords. Consider factors such as connector styles, fiber types, cable construction, and insertion loss.

12.2 Using vendor data sheets, list some characteristics of a typical fiber distribution frame (FDF) and a fiber entrance cabinet. For the FDF consider factors such as fiber termination capacity, physical size, frame access, and connector type. For the entrance cabinet list parameters such as splice capacity, fiber management process (including bend radius protection), cable access, and physical size.

12.3 As noted in Section 12.1, an important factor for an indoor cable, such as a patch cord, is the *flammability rating*. List details of the various test procedures and acceptance criteria established by Underwriters' Laboratories (UL) for indoor cable flammability compliance (e.g., consider UL 1581, UL 1666, and UL 1685).

12.4 Using vendor data sheets, list some characteristics of typical distribution cabinets and access terminals that can be used in the outside plant for FTTP networks.

12.5 Two 5-m-high telephone poles are separated by 45 m. How high off the ground is the center of the cable if the sag allowed is 1 percent of the span length between poles?

12.6 When cables are installed in ducts, the ratio of the sum of the cross-sectional areas of the cables to the inside cross-sectional area of the duct is called the *fill ratio*. Suppose that three optical fiber cables which have outer diameters of 6.0, 7.5, and 9.0 mm are blown into a duct that has a 20-mm inner diameter. What is the fill ratio for this case?

12.7 Make a diagram of a complete FTTP network cable layout from the OLT and video equipment in the central office to an ONT at a subscriber site. Show all connectors, splices, and interface boxes along the path of a fiber.

12.8 Examine the details of how to implement horizontal drilling in several different situations: for example, across a garden area, under a road or driveway, and under a stream. Possible factors to consider are what size machine is needed, how long the installation process takes, what types of conduits are installed, and how to handle different types of soil conditions.

12.9 Examine the details of how to implement an air-assisted optical fiber cable installation for a drop cable to a user's premises and for a 2-km distribution cable in an underground duct. Possible factors to consider are what size machine is needed for various cable sizes, how long the installation process takes, and how to prevent damage to the cable during installation.

12.10 A large percentage of FTTP cabling in the outside plant consists of aerial installations. Write a short report describing the general installation procedures for running aerial fiber optic distribution and drop cables from an aboveground distribution cabinet to a customer site (e.g., see Figure 12.9). *Hint*: You may want to look at specifications for aerial cable installations on the Web sites of optical cable vendors (e.g., www.corningcablesystems.com).

12.11 Show where various types of labels and warning markers might be placed along a typical cable route from an OLT in the central office to an ONT at a subscriber site. What kind of information should each label or marker contain?

FURTHER READING

1. Bob Chomycz, *Fiber Optic Installer's Field Manual*, McGraw-Hill, New York, 2000.
2. BiCSI, *Telecommunications Cabling Installation*, McGraw-Hill, New York, 2001.
3. A. L. Crandall, C. R. Herron, and R. B. Washburn, "Controlling axial load forces on optical fiber cables during installation," *OFC/NFOEC 2005 Conf. Program*, Anaheim, CA, Mar. 6–11, 2005, Paper NWC3.
4. G. Keiser, *Optical Communications Essentials*, McGraw-Hill, New York, 2003.
5. O. L. Storaasli, "Compatibility of fiber optic microduct cables with various blowing installation equipment," *OFC/NFOEC 2005 Conf. Program*, Anaheim, CA, Mar. 6–11, 2005, Paper NWC2.
6. O. L. Storaasli, B. Overton, and R. Lovie, "FTTU drop cable reliability and applications considerations," *OFC/NFOEC 2005 Conf. Program*, Anaheim, CA, Mar. 6–11, 2005, Paper NWL3.
7. TIA/EIA-590-A, "Standard for Physical Location and Protection of Below-Ground Fiber Optic Cable Plant," July 2001.
8. M. D. Vaughn, D. Kozischek, D. Meis, A. Boskovic, and R. E. Wagner, "Value of reach-and-split ratio increase in FTTP access networks," *J. Lightwave Technol.*, vol. 22, pp. 2617–2622, Nov. 2004.

13

NETWORK INSTALLATION TESTING

The deployment, operation, and maintenance of a PON require measurement techniques for verifying that the network has been configured properly and that its constituent components are functioning correctly. Prior to the widespread installation of FTTP networks, most measurements involved checking the operational status of point-to-point links for long-haul or metro applications. The use of PON technology introduces a new challenge to network testing, since there is a passive optical splitter in the outside cable plant. Now the network becomes a point-to-multipoint architecture that requires enhanced test and measurement instruments. In addition, these instruments must be capable of measuring the performance of a single bidirectional fiber link that carries three wavelengths simultaneously.

The main parameters that test instruments need to measure in an FTTP network are the optical power level at a variety of network points, loss of power as light passes through the various PON elements, and optical return loss. The appropriate instruments include optical power meters, light sources emitting at the three specific FTTP wavelengths (1310, 1490, and 1550 nm), visual fault indicators, optical time-domain reflectometers, and special optical-loss test sets.

Of particular importance are accurate and precise performance characterizations of the optical fibers in the different outside plant links. Since there is an optical power splitter in the transmission path, upstream and downstream measurements may need to done differently. In addition, bidirectional measurements must be made at the three FTTP wavelengths. Following installation, various test methods are needed to monitor the link condition continually to verify that the performance requirements are met during operation. Other measurements relate to network maintenance, such as locating breaks or faults in optical fiber cables and checking the status of backup batteries.

FTTX Concepts and Applications, by Gerd Keiser
Copyright © 2006 John Wiley & Sons, Inc.

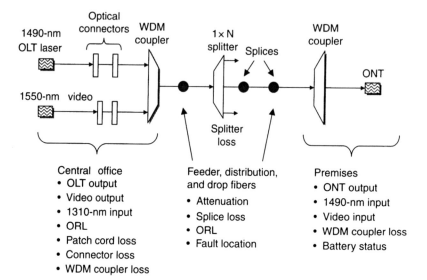

Figure 13.1. Points in an FTTP network where some of the relevant test parameters are of importance.

In this chapter we discuss performance measurements and operational tests of interest to installers, operators, and repair personnel of FTTP networks. Figure 13.1 shows some of the relevant test parameters and at what points in an FTTP network they are of importance. The discussion here focuses on the optical characteristics of a PON. We first address measurement standards in Section 13.1 and basic test equipment for passive optical network links in Section 13.2. Next, optical power and its measurement with optical power meters that have been optimized for FTTP network applications are discussed in Section 13.3.

In Section 13.4 we describe the characteristics and use of an optical time-domain reflectometer, which is a versatile instrument that is used widely to evaluate the characteristics of an installed optical fiber link. An important parameter to measure in a PON is the optical return loss, which is the percent of total reverse power in relation to total forward power at points such as connectors, fiber ends, optical splitter interfaces, and within the fiber itself due to Rayleigh scattering. This effect and the instrument used to measure it are described in Section 13.5. Another useful tool is a visual fault locator. In Section 13.6 we describe this handheld pen-size instrument that uses a visible laser light source to locate anomalies such as fiber breaks, overly tight bends in a cable, or poorly mated connectors. Any such point in an optical link is referred to as an *event*. An important instrument described in Section 13.7 is the optical-loss test set, which is used for making fiber loss and length measurements in an optical network link. Instead of having a truck full of different types of test equipment, installation and maintenance personnel now can use multifunction test equipment. An overview of the measurement functions such an instrument can carry out is given in Section 13.8. Section 13.9 covers another specialized instrument that has been designed for device conformance verification.

A series of optical tests should be done during the deployment of a PON to verify that the cable plant and the transmission equipment at either end will meet the design specifications. In Section 13.10 we give details on the basic measurement procedures and describe which instruments can be used for various tests.

13.1 INTERNATIONAL MEASUREMENT STANDARDS

As summarized in Table 13.1, several different international organizations are involved in formulating equipment and testing standards. The major organizations that deal with measurement methods for links and networks are the Institute for Electrical and Electronic Engineers (IEEE) and the Telecommunication Standardization Sector of the International Telecommunication Union (ITU-T). Telcordia Technologies provides a wide range of generic requirements for telecommunications network components and equipment. Appendix E gives some details on Telcordia generic requirements for FTTP applications.

A key organization for component testing is the Telecommunication Industry Association (TIA) in association with the Electronic Industries Alliance (EIA). The TIA has a list of over 120 fiber optic test standards and specifications under the general designation TIA/EIA-455-XX-YY, where XX refers to a specific measurement technique and YY refers to the publication year. These standards are also called *Fiber Optic Test Procedures* (FOTP), so that TIA/EIA-455-XX becomes FOTP-XX.

TABLE 13.1 Summary of Key Standards Organizations and Their Functions Related to PON Testing

Organization	Internet Address	PON-Related Activities
IEEE	www.ieee.org	Establish and publish measurement procedures for links and networks • Define physical-layer test methods • IEEE 802.3ah Ethernet in First Mile (EFM)
ITU-T	www.itu.int/ITU-T	Create and publish standards in all areas of telecommunications • Series G for telecommunications • Series L for outside plant elements • ITU-T Recommendation G.983 (BPON) • ITU-T Recommendation G.984 (GPON) • ITU-T Recommendation G.985 (EPON)
Telcordia	www.telcordia.com	Provide generic requirements for network elements • Fiber optic connectors • Indoor and outdoor cabinets • Underground, aboveground enclosures • Field-deployed products
TIA/EIA	www.tiaonline.org www.eia.org	Created over 120 test specifications under the designation "Fiber Optic Test Procedures" • Define physical layer test methods • TIA/EIA-455-XX or FOTP-XX documents

These include a wide selection of recommended methods for testing the response of fibers, cables, passive devices, and electrooptic components to environmental factors and operational conditions. For example, TIA/EIA-455-20-B-2004, or FOTP-20, is a method published in 2004 that gives two procedures for monitoring the changes in transmittance of optical fibers or cables that may occur during mechanical or environmental testing.

13.2 BASIC TEST INSTRUMENTS

The basic measurement and test instruments for FTTP networks include optical power meters, light sources emitting at the three specific FTTP wavelengths (1310, 1490, and 1550 nm), visual fault indicators, optical time-domain reflectometers, and special optical return loss testers. These instruments enable a variety of statistical measurements to be made at the push of a button, after the user has keyed in the parameters to be tested and the desired measurement range.

The equipment is available with a selection of capabilities. Its sizes range from portable, handheld units for field use to sophisticated briefcase-sized benchtop or rack-mountable instruments for factory or central office applications. In general, the field units do not need to have the extremely high precision of laboratory instruments, but they need to be more rugged to maintain reliable and accurate measurements under extreme environmental conditions of temperature, humidity, dust, and mechanical stress. However, even the handheld equipment for field use has reached a high degree of sophistication with automated microprocessor-controlled test features and laptop computer interface capabilities.

Table 13.2 lists some test instruments used for the deployment, operation, and maintenance of a PON. Measurement methodologies for analyzing the performance

TABLE 13.2 Some Widely Used Optical System Test Instruments for PONs and Their Functions

Test Instrument	Function
Optical power meter	Measures optical power over a selected wavelength band
Visual fault indicator	Uses visible light to give a quick indication of a break in an optical fiber
Test-support lasers (multiple-wavelength or broadband)	Assist in tests that measure the wavelength-dependent response of an optical component or link
Optical power attenuator	Reduces power level to prevent instrument damage or to avoid overload distortion in the measurements
OTDR (field instrument)	Measures attenuation, length, connector/splice losses, and reflectance levels; helps locate fiber breaks
Multifunction optical test system	Factory or field instruments with exchangeable modules for performing a variety of measurements
Optical return loss tester	Measures total reverse power in relation to total forward power at a particular point
Optical spectrum analyzer (OSA)	Measures optical power as a function of wavelength
BER test equipment	Uses standard eye-pattern masks to evaluate the data-handling ability of an optical link

characteristics of optical fibers and various passive and active components are not covered here but may be found in the book by Derickson.

13.3 OPTICAL POWER MEASUREMENTS

Optical power measurement is the most basic function in fiber optic metrology. This parameter is not a fixed quantity and can vary as a function of parameters such as time, distance along a link, wavelength, phase, and polarization.

13.3.1 Definition of Optical Power

To get an understanding of optical power, let us consider its physical basis.

- Light particles, called *photons*, have a certain energy associated with them, which changes with wavelength. The relationship between the energy E of a photon and its wavelength λ is given by the equation $E = hc/\lambda$, which is known as *Planck's law*. Here $c = 3 \times 10^8$ m/s (299,792,458 m/s) is the speed of light and the parameter $h = 6.63 \times 10^{-34}$ J·s $= 4.14$ eV·s is called *Planck's constant*. In terms of wavelength (measured in units of µm), the energy in electron volts is given by the expression E (eV) $= 1.2406/\lambda$ (µm). Note that 1 eV $= 1.60218 \times 10^{-19}$ J.
- *Optical power* P measures the rate at which photons arrive at a detector. Thus, it is a measure of energy transfer per time. Since the rate of energy transfer varies with time, the optical power is a function of time. It is measured in *watts* or joules per second (J/s).

Since optical power varies with time, its measurement also changes with time. Figure 13.2 shows plots of the power level in a signal pulse as a function of time.

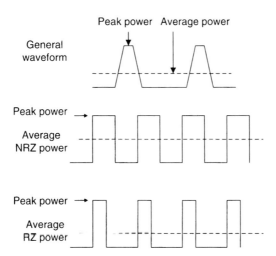

Figure 13.2. Peak and average powers in a series of general, NRZ, and RZ optical pulses.

It is clear that different instantaneous power-level readings are obtained depending on the exact instance when the measurement is made. Therefore, two standard classes of power measurements can be specified in an optical system. These are the peak power and the average power. The *peak power* is the maximum power level in a pulse, which might be sustained for only a very short time.

The *average power* is a measure of the power level averaged over a relatively long time period compared to the duration of an individual pulse. For example, a measurement time period of 0.1 second for a 155-Mbps data stream contains millions of signal pulses. As a simple example, in a non-return-to-zero (NRZ) data stream (see Appendix D) there will be an equal probability of 1 and 0 pulses over a long time period. In this case, as shown in Figure 13.2, the average power is half the peak power. If a return-to-zero (RZ) modulation format is used, the average power over a long sequence of pulses will be one-fourth the peak power since there is no pulse in a 0 time slot and a 1 time slot is only half-filled.

The *sensitivity of a photodetector* normally is expressed in terms of the average optical power level impinging on it, since the measurements in an actual fiber optic system are done over many pulses. However, the output level for an optical transmitter normally is specified as the peak power. This means that the average power coupled into a fiber, and the power level that a photodetector measures, is at least 3 dB lower than if the link designer uses the peak source output inadvertently in power-budget calculations as the light level entering the fiber.

13.3.2 Optical Power Meter

The function of an *optical power meter* is to measure power over a selected spectral passband. For example, for FTTP applications an optical power meter might have the following passbands: 1260 to 1360 nm for measurements at 1310 nm, 1480 to 1500 nm for measurements at 1490 nm, and 1540 to 1565 nm for measurements at 1550 nm. Handheld instruments for these applications are used extensively during all phases of FTTP network deployment, operation, and maintenance. These instruments come in a wide variety of types with different levels of capabilities. Multiwavelength optical power meters using photodetectors are the most common instrument for measuring optical signal power levels. Usually, the meter outputs are given in dBm (where 0 dBm = 1 mW) or dBµ (where 0 dBµ = 1 µW).

Figure 13.3 shows an example of a handheld fiber optic power meter that has been designed for FTTP service activation and maintenance. A special feature of this meter is that it can act as a pass-through device, which means that it can be connected between an operating OLT and an ONT and not interfere with the live voice, data, and video traffic on the link. In this configuration, the meter extracts a small percentage of the signal power for use by the photodetectors in the power meter. This allows a simultaneous measurement and display of the optical power at all the three standard FTTP wavelengths. The meter has 10 threshold settings and pass–fail indicators to see whether or not all the voice, data, and video signals fall within their specified ranges. The ability to set such thresholds enables the meter to be used for

OPTICAL POWER MEASUREMENTS 239

Figure 13.3. Handheld fiber optic power meter designed for FTTP service activation and maintenance. (Model PPM-350B, photo provided courtesy of EXFO; www.exfo.com.)

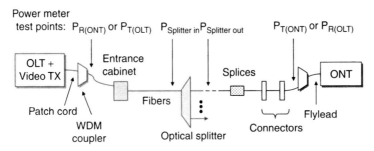

Figure 13.4. User-selected threshold settings allow an optical power meter to make quick performance checks for network setup or troubleshooting at any point in a PON.

network setup or troubleshooting at any point from the OLT to the ONT, as Figure 13.4 shows. The power measures could include the following:

- Transmitted signal levels $P_{T(ONT)}$ from an ONT or $P_{T(OLT)}$ from an OLT
- Received signal levels $P_{R(ONT)}$ from an ONT or $P_{R(OLT)}$ from an OLT
- Power levels $P_{\text{splitter in}}$ or $P_{\text{splitter out}}$ at the splitter input and output ports, respectively

13.3.3 Power Meter Applications

Instruments such as the one shown in Figure 13.3 will display the optical power-level measurements at all three FTTP wavelengths and will give a pass, fail, or warning

1310 nm ONT	−4.9 dBm	Warning
1490 nm OLT	−21.1 dBm	Pass
1550 nm VIDEO	0.9 dBm	Pass

Figure 13.5. Optical power meter display of measurement values made at an ONT. Signals arriving at the ONT have adequate levels, but the ONT optical output is marginal.

indicator based on the thresholds set for that particular network point. For example, a measurement made at an ONT might display the information shown in Figure 13.5. This indicates that the voice, data, and video signals arriving at the ONT from the OLT have adequate levels but that the ONT output in the upstream direction is marginal.

13.4 OPTICAL TIME-DOMAIN REFLECTOMETER

An *optical time-domain reflectometer* (OTDR) is a versatile instrument that is used widely to evaluate the characteristics of an installed optical fiber link. In addition to identifying and locating faults or anomalies within a link, this instrument measures parameters such as fiber attenuation, length, optical connector and splice losses, and light reflectance levels.

An OTDR is fundamentally an optical radar. It operates by launching narrow laser pulses periodically into one end of a fiber under test by using either a directional coupler or a beam splitter. The properties of the optical fiber link are then determined by analyzing the amplitude and temporal characteristics of the waveform of the backscattered light. A typical OTDR consists of a light source and receiver, data acquisition and processing modules, an information-storage unit for retaining data either in the internal memory or on an external disk, and a display. Figure 13.6 shows a portable OTDR that can be used for making measurements in the field.

13.4.1 OTDR Trace

Figure 13.7 shows a typical trace as seen on the display screen of an OTDR. The scale of the vertical axis is logarithmic and measures the returning (back-reflected) signal in decibels. The horizontal axis denotes the distance between the instrument and the measurement point in the fiber. In addition to the trace, an OTDR such as the one shown in Figure 13.7 also can place a number next to an event on the display

OPTICAL TIME-DOMAIN REFLECTOMETER   241

Figure 13.6. Portable OTDR for making measurements in the field. (Model FTB-400, photo provided courtesy of EXFO; www.exfo.com.)

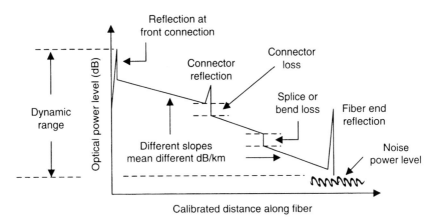

Figure 13.7. Representative trace of backscattered and reflected optical power as displayed on an OTDR screen and the meanings of various trace features.

and give a list of these numbers and their corresponding measurement information in a table below the trace.

The backscattered waveform has four distinct features:

- A large initial pulse resulting from Fresnel reflection at the input end of the fiber.
- A long decaying tail resulting from Rayleigh scattering in the reverse direction as the input pulse travels along the fiber.
- Abrupt shifts in the curve caused by optical loss at joints or connectors in the fiber line.

- Positive spikes arising from Fresnel reflection at the far end of the fiber, at fiber joints, and at fiber imperfections.

Fresnel reflection and Rayleigh scattering principally produce the backscattered light. *Fresnel reflection* occurs when light enters a medium having a different index of refraction. For a glass–air interface, when light of power P_0 is incident perpendicular to the interface, the reflected power P_{ref} is

$$P_{ref} = P_0 \left(\frac{n_{fiber} - n_{air}}{n_{fiber} + n_{air}} \right)^2 \tag{13.1}$$

where n_{fiber} and n_{air} are the refractive indices of the fiber core and air, respectively. A perfect fiber end reflects about 4 percent of the power incident on it. However, since fiber ends generally are not polished perfectly and perpendicular to the fiber axis, the reflected power tends to be much lower than the maximum possible value. In particular, this is the case if an *angle-polished connector* (APC) is used.

The detection and measurement accuracy of an event depend on the signal-to-noise ratio (SNR) that an OTDR can achieve at that point. This is defined as the ratio between the back-reflected signal and the noise level. The SNR depends on factors such as the OTDR pulse width, how often the OTDR samples the signal, and the distance to the measurement point.

Two important performance parameters of an OTDR are dynamic range and measurement range. *Dynamic range* is defined as the difference between the initial backscatter power level at the front connector and the noise-level peak at the far end of the fiber (see Figure 13.7). It is expressed in decibels of one-way fiber loss. Dynamic range provides information on the maximum fiber loss that can be measured and denotes the time required to measure a given fiber loss. A basic limitation of an OTDR is the trade-off between dynamic range and *event location resolution*. For high spatial resolution, the pulse width has to be as small as possible. However, this reduces the signal-to-noise ratio and thus lowers the dynamic range. Typical distance resolution values range from 8 cm for a 10-ns pulse to 5 m for a 50-µs pulse.

Measurement range deals with how far away an OTDR can identify events in the link, such as splice points, connection points, or fiber breaks. The maximum range R_{max} depends on the fiber attenuation α and on the pulse width, that is, on the dynamic range D_{OTDR}. If the attenuation is given in dB/km, the maximum range in kilometers is

$$R_{max} = \frac{D_{OTDR}}{\alpha} \tag{13.2}$$

For example, if the dynamic range is 36 dB and the attenuation is 0.5 dB/km, the maximum range is 72 km.

13.4.2 OTDR Dead Zone

The concept of a dead zone is another important OTDR specification. *Dead zone* is the distance over which the photodetector in an OTDR is saturated momentarily after it measures a strong reflection. As Figure 13.8 shows, there are two specifications for

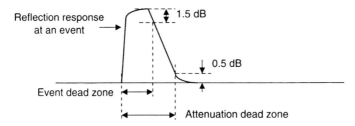

Figure 13.8. Two specifications for dead zone are the event dead zone and the attenuation dead zone.

dead zone. An *event dead zone* specifies the minimum distance over which an OTDR can detect a reflective event that follows another reflective event. Typically, vendors specify this as the distance between the start of a reflection and the -1.5-dB point on the falling edge of the reflection. A short pulse width is used when measuring the event dead zone. For example, a 30-ns pulse width would give a 3-m event dead zone.

The *attenuation dead zone* indicates over which distance the photodetector in an OTDR needs to recover following a reflective event before it is again able to detect a splice. This means that the receiver has to recover to within 0.5 dB of the backscatter value. Typical attenuation dead zones range from 10 to 25 m.

Typically, an OTDR dead zone is the same length as the distance that the optical pulse covers in a fiber plus a few meters. Thus OTDR vendors have started employing a special length of fiber called an *optical pulse suppressor* (OPS). An OPS moves the dead zone from the beginning of the fiber under test to this special fiber. This can reduce the event dead zone to about 1 m, so that anomalies occurring within a short distance, for example within the cabling system of a central office, may be detected and measured.

13.4.3 Fiber Fault Location

To locate breaks and imperfections in an optical fiber, the fiber length L (and hence the position of the break or fault) can be calculated from the time difference between the pulses reflected from the front of the fiber and the event location. If this time difference is t, the length L is given by

$$L = \frac{ct}{2n_1} \tag{13.3}$$

where n_1 is the core refractive index of the fiber. The number "2" in the denominator accounts for the fact that light travels a length L from the source to the break point and then another length L on the return trip.

13.5 OPTICAL RETURN LOSS

Reflections of light in a backward direction occur at various points in optical links that use laser transmitters. This can occur at connectors, fiber ends, optical splitter

Figure 13.9. Compact, portable optical return loss (ORL) tester, which can also be used as a power meter. (Model 7340, photo provided courtesy of Kingfisher; www.kingfisher.com.au.)

interfaces, and within the fiber itself due to Rayleigh scattering. The percent of power reflected back from a particular point in a lightpath is called *back reflection*. If they are not controlled, back reflections can cause optical resonance in the laser source and result in erratic operation and increased laser noise. In addition, back reflections can undergo multiple reflections in the transmission line and increase the bit error rate when they enter the receiver.

It is important, therefore, to measure the *optical return loss* (ORL), which is the percent of total reverse power in relation to total forward power at a particular point. The ORL is expressed as a ratio of reflected power P_{ref} to incident power P_{inc}:

$$\text{ORL} = 10 \log \frac{P_{ref}}{P_{inc}} \tag{13.4}$$

One can use either an OTDR or an ORL meter to measure this parameter. Although an OTDR can give precise reflectance values at individual events along a fiber transmission path, it has a limitation in measuring the back reflections near and within the OTDR dead zone. Since such an event can be a major contributor to ORL, it is better to use a return loss meter, such as the one shown in Figure 13.9.

13.6 VISUAL FAULT LOCATOR

A *visual fault locator* (VFL) is a handheld pen-sized instrument that uses a visible laser light source to locate events such as fiber breaks, overly tight bends in a fiber, or poorly mated connectors. The source emits a bright beam of red light (e.g., 650 nm) into a fiber, thereby allowing the user to see a fiber fault or a high-loss point as a glowing or blinking red light. It is particularly useful for identifying fiber faults within the initial dead zone of an OTDR. In using such a device, events must occur where the fiber or connector is in the open so that visual observation of the emitted red light is possible.

The nominal light output is 1 mW, so the light will be visible through a fiber jacket at a fault point. This power level allows a user to detect a fiber fault visually for up to 5 km. A number of companies produce such a tool. The device generally is powered by one 1.5-V size AA battery and operates in either a continuously on or a blinking mode.

13.7 OPTICAL-LOSS TEST SET

An *optical-loss test set* (OLTS) is an instrument for making fiber loss and length measurements in an optical network link. By using a light source and an optical power meter an OLTS measures fiber loss directly by computing the difference between the optical power entering a fiber and the optical power exiting it. A major application is during the installation, provisioning, and troubleshooting of optical fiber cable plants for telecommunications and FTTP networks.

Some units can perform fully automated bidirectional loss tests and may include optional ORL, VFL, and optical talk-set functions. The optical talk set allows full-duplex communications between technical personal located at opposite ends of the link being tested.

13.8 MULTIFUNCTION TEST INSTRUMENT

Instead of having a truck full of different types of test equipment, various manufacturers are producing multifunction test equipment. Figure 13.10 shows one example from a number of available instruments. This particular model has been engineered for FTTP use and combines the functions of the following eight instruments into one unit:

- Loss meter
- Power meter
- Optical return loss meter
- Visual fault locator
- Multimode and single-mode light sources
- Digital talk set
- Fiber-length tester
- Video fiber inspection probe

The unit also contains optical sources to carry out more sophisticated optical power measuring at the recommended 1310-, 1490-, and 1550-nm wavelengths for

Figure 13.10. Compact, portable multifunction test instrument for use in field environments. (Model FOT-930, photo provided courtesy of EXFO; www.exfo.com.)

PONs. Thus, a single portable instrument can function as a power meter, an optical-loss tester for measuring loss in a fiber automatically in two directions at two wavelengths, an optical-return-loss tester, a device for measuring the quality of optical patch cords, a visual fault indicator for locating breaks and failures in a fiber cable, and a talk set for full-duplex communications between field personnel.

13.9 DEVICE CONFORMANCE TESTING

To fully check the operational characteristics of a device prior to its installation, it is best to examine its performance under a variety of calibrated optical signal degradations. This procedure is known as *conformance testing*. Figure 13.11 shows a specialized optical standards tester (OST) which has been designed for that purpose. With such an instrument, an engineer can examine the response of a device by degrading a perfect test signal to make it correspond more to an imperfect real-world signal. This includes varying the optical power level, extinction ratio, phase and amplitude jitter, interfering power from another laser, and the optical signal-to-noise ratio (OSNR). These parameters can be changed independently or in combination with one another.

The instrument can carry out and display performance measure parameters and characteristics such as the bit error rate, how optical noise affects the *device under test* (DUT), the receiver optical sensitivity, at what power level a receiver overloads

Figure 13.11. Specialized optical standards tester (OST) designed for conformance testing. (Photo provided courtesy of Circadiant Systems; www.circadiant.com.)

and starts generating errors, and the path penalty resulting from using the DUT. As Figure 13.11 illustrates, as part of its data storage and comparison capabilities, the instrument can scan a barcode identifier for each component to include with the test data.

13.10 FTTP NETWORK TESTING

A number of optical tests should be done during the deployment of a PON to verify that the cable plant and the transmission equipment at either end will meet the design specifications. In this section we describe the basic measurement procedures and describe which instruments can be used for these tests. References 4 and 6 present greater details on measurements of specific network cabling examples.

Figure 13.12 shows the general measurements to be made between the OLT in the central office and the ONT at the customer's premises. These include:

- *Test 1:* characterizing the individual links both in the central office and in the outside cable plant
- *Test 2:* verifying that the total bidirectional end-to-end loss budget is satisfied
- *Test 3:* measuring the losses of fiber splices, optical connectors, and wavelength couplers
- *Test 4:* checking the bidirectional end-to-end ORL
- *Test 5:* verifying that the ONT receives the proper optical signal levels from the OLT and from the video equipment in the central office (1490- and 1550-nm wavelengths)
- *Test 6:* verifying that the OLT receives the proper 1310-nm optical signal levels from each ONT

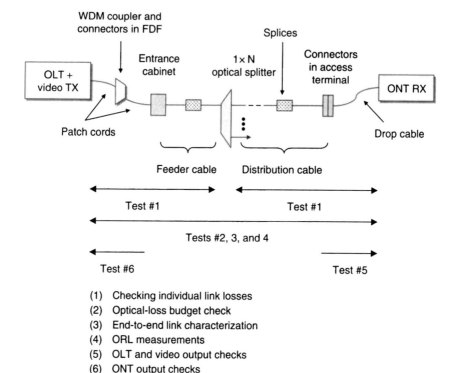

Figure 13.12. General measurements between the OLT in the central office and the ONT at the customer's premises.

13.10.1 Checking Individual Link Losses

The loss of individual fibers in the entire network should be checked before they are attached to the power splitter. This needs to be done bidirectionally and includes both the feeder fiber and the distribution and drop cables running between the splitter and the ONTs. The downstream measurements need to be done at 1490 and 1550 nm, since the attenuation varies with wavelength. Similarly, in the upstream direction the loss should be measured at 1310 nm. These bidirectional checks can be done in several ways, depending on available equipment. All methods require some type of coordination and communication between test personnel at both ends of a link. Test setup examples include:

1. A calibrated light source on one end and an optical power meter at the other end.
2. An OLTS at each end. Each OLTS has an integrated light source and power meter.
3. A pair of multifunction testers such as shown in Figure 13.10 in place of the OLTS tools.

The second and third setups are more desirable, particularly if each instrument can perform automated testing and has a built-in triple-wavelength capability. These features reduce testing time and minimize the possibility of operator errors. Advanced OLTS units and multifunction testers also have the ability to produce test result tables automatically, including average loss and worst-case optical return loss. Thus, these instruments can generate and store complete measurement reports at both locations.

13.10.2 Optical-Loss Budget Check

As described in Chapters 6 through 9, each type of PON can be designed to either a class B or C loss budget of 25 or 30 dB, respectively. Typically, the lower-level class A budget design of 20 dB is not used, since this puts a severe strain on the component specifications.

As shown in Figure 13.13, after the feeder and distribution fibers have been attached to the optical splitter and the central office patch cords are in place, the total loss between the OLT and the ONT can be determined with a pair of optical-loss test sets. Here there is an OLTS at each end of the path under test, one unit being the master and the other the slave. By using a laser diode light source built into the OLTS, one can measure optical losses up to 67 dB. If the measured value exceeds the loss budget, error-free transmission is not achievable. The downstream measurements need to be done at 1490 and 1550 nm, since the attenuation varies with wavelength. Similarly, in the upstream direction the loss should be measured at 1310 nm. The bidirectional checks should be done for each branch of the optical splitter, that is, between each ONT termination and the OLT connection point in the central office.

13.10.3 End-to-End Link Characterization

An OTDR can give a detailed overall picture of the end-to-end link characteristics. This can be done from just one end. It includes parameters such as the attenuation

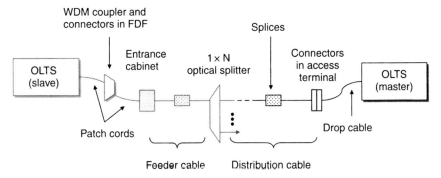

Figure 13.13. Determining the total loss between the OLT and the ONT with a pair of optical-loss test sets.

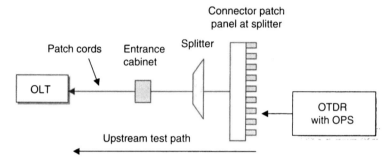

Figure 13.14. Upstream setup for checking losses between the splitter outputs and the OLT-to-patch cord interface.

of individual fiber segments; the location and losses of splices, connectors, optical splitters, and WDM couplers; and link anomalies. The anomalies can come from overly tight bends in the fiber (e.g., in equipment cabinets), possible cracks, and mismatches in fiber core size.

Figure 13.14 shows a setup for checking optical splitter, WDM coupler, and optical connector losses from the far downstream ports of the splitter to the OLT-to-patch cord interface. Here a special *optical pulse suppressor* (OPS) device is placed between the OTDR and the splitter. An OPS consists of a fiber coil ranging in size from 300 to 3000 m which is used to mask the dead zone of the OTDR.

Besides the upstream test, the downstream end-to-end link characteristics from the OLT-to-patch cord interface to the far-downstream ports at the ONT termination should be checked. To do this with an OTDR, each leg of the distribution segment from the optical splitter to an ONT endpoint must have a slightly different length. This will result in distinct events for each splitter output leg being shown clearly on the OTDR display screen. If all legs had the same length, their backscattered and reflected light signal traces would all fall on top of each other on the OTDR display.

A basic purpose of performing bidirectional tests is to check for the possible occurrence of factors such as directional losses through connectors and core mismatches. Core mismatches arise from geometric differences in the cores of mated fibers, particularly in patch cords in the central office. These mismatches can be due to variations in the core diameter, ellipticity, or concentricity of two mated fibers. Of special concern is a mismatch in the core sizes, which can lead to high losses in one direction but not in the other. To see this effect for two different fibers, let the subscripts E and R refer to the emitting and receiving fibers, respectively. If all characteristics of the two fibers are the same except for their fiber diameters d_E and d_R, the coupling loss L_F due to core mismatches is

$$L_F(d) = \begin{cases} -10 \log\left(\dfrac{d_R}{d_E}\right)^2 & \text{for } d_R < d_E \\ 0 & \text{for } d_R \geq d_E \end{cases} \quad (13.5)$$

Let us look at some examples. First consider the case when trying to connect a 62.5-μm fiber to one with a 50-μm core. Light traveling from the larger to the smaller fiber

TABLE 13.3 Optical Power Losses When Light Goes from a Larger to a Smaller Fiber

Emitting Diameter[a] (μm)	Receiving Diameter[a] (μm)	Loss (dB)
62.5 (MM)	50 (MM)	1.9 (36%)
62.5 (MM)	9 (SM)	16.8 (98%)
9.8 (SM)	8.8 (SM)	0.94 (20%)
9.5 (SM)	9.1 (SM)	0.37 (8%)

[a]MM and SM designate multimode and single-mode fibers, respectively.

experiences a 1.9-dB loss, or 36 percent of the power. However, there is no loss in the opposite direction. A much more serious loss occurs when one inadvertently tries to couple light from a multimode to a single-mode fiber. For example, suppose that someone connects a 62.5-μm multimode fiber to a 9-μm single-mode fiber. This mismatch results in an area-mismatch loss of 17 dB, or almost 98 percent for light traveling from the larger to the smaller fiber. However, there is no loss for light traveling from the smaller to the larger fiber. A bidirectional link evaluation will spot such inadvertent connection mismatches. Table 13.3 illustrates some possible losses in going from a larger to a smaller fiber. Note that a relatively large loss can occur at a fiber-to-fiber joint even when connecting two single-mode fibers if their geometric properties are different.

13.10.4 ORL Measurements

Bidirectional ORL measurements can be made either with a pair of ORL meters or multifunction testers at each end of the complete OLT-to-ONT link. This reflectance testing should be done after the OLT-to-ONT link is completed. For a class B passive optical network the reflected optical power should be at least 30 dB below the incident power level. Generally, a value in the range −30 to −35 dB is acceptable. However, values less than −30 dB should trigger corrective action.

13.10.5 OLT and Video Output Checks

After the network characterization is completed on all links, the installation personnel should check the optical power levels arriving at the drop point of each ONT from the central office equipment. This can be done using a power meter that in addition to providing a power-level reading can be set to a specific pass–fail threshold. The power level should be checked for both 1490 nm (the OLT voice and data output) and 1550 nm (the video signal level). These procedures are necessary to verify that sufficient optical power arrives at each ONT. The OLT output should be checked at several points in the network. These points include:

- The exit port of the WDM coupler in the central office
- Each exit port of the optical splitter
- The end of each drop cable at an ONT

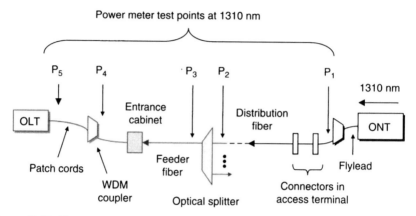

Figure 13.15. Measuring upstream optical power levels from an ONT at various points in the network.

13.10.6 ONT Output Check

To verify that the network evaluation was done correctly, installation personnel should check the 1310-nm upstream optical power levels arriving at various points in the network from each ONT. To do these measurements requires an optical power meter with a *pass-through testing capability* (such as the model illustrated in Figure 13.3). This capability means that when it is connected in-line between the OLT and an ONT, the meter extracts a small percentage of the signal power for use by its photodetector. The reason for this special function is that an ONT only sends out signals in response to transmission permits from the OLT. Thus, the PON link must be operating in order to measure upstream power levels emitted by the ONT.

As illustrated in Figure 13.15 for upstream testing, first the output P_1 of the ONT can be measured with an optical power meter to verify that it is operating properly. Next, the upstream output P_2 exiting the distribution cable can be measured at the interface junction to the optical splitter. Obviously, this only can be done if the distribution cable-to-splitter interface is made using an optical connector, which is the method preferred to that of splicing the distribution fiber to the splitter. Similarly, P_3 is the upstream output from the splitter at the interface to the feeder cable. Finally, levels P_4 at the fiber distribution frame (FDF) and P_5 after the final patch cord can be checked in the central office. Unless a special signal-monitoring connection has been designed into the system (e.g., within the fiber distribution frame), this can be done only before the network is put into actual service.

13.11 FTTP NETWORK TROUBLESHOOTING

After an FTTP network is installed and put into operation, the inevitable electronic and optical fiber cable plant degradations and failures will occur. Among the long list of operational anomalies or failures that might occur at an ONT, at the OLT, or in the OSP are the following:

- OLT or ONT circuit card failures
- Degraded optical connectors resulting from moisture, dirt, damage, or misalignment
- Performance degradation in an ONT resulting from other customer-connected equipment
- Backup battery or power supply failures
- Malicious or accidental damage to a distribution cabinet or access terminal
- Cable cuts caused by errant backhoes, storm damage, or gnawing animals

These conditions can lead to loss of optical signal at one or more ONTs, received optical power that is below its specified value, an increased BER, or a degraded signal. Thus, there always will be the need for ongoing maintenance and troubleshooting of the network. Software-based system diagnostics tools can identify and locate many of these problems and can assign appropriate action for their resolution. A question then arises as to what procedures a service technician must follow to resolve various types of faults.

13.11.1 Resolutions of Network Problems

To get an appreciation of network troubleshooting tasks, let us consider a few scenarios of possible locations of operational problems and what steps can be taken for their resolution. Following the lead of the *FTTx PON Guide* from EXFO (see Reference 6), for ease in identifying where a fault occurs, it is convenient to divide the link path between the OLT and an ONT into seven troubleshooting zones. Figure 13.16 shows the ranges of these seven zones.

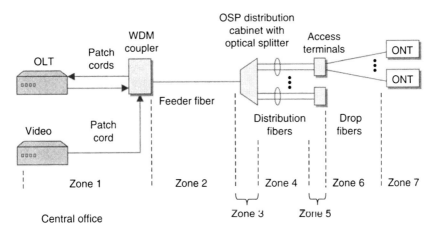

Figure 13.16. Definition of PON troubleshooting zones.

First, suppose that all ONTs on the network lose either some or all of their services. In this case the fault lies between the electrical input to the OLT and the light output of the optical splitter. Possible trouble origins could be a malfunctioning OLT or video transmitter (zone 1), broken or disconnected patch cords in the CO (zone 1), a fault along the feeder cable (zone 2), or the optical splitter in zone 3 could be damaged (e.g., a vehicle may have crashed into the distribution cabinet).

The easiest troubleshooting procedure is to start in zone 1 at the central office and check the status of the transmission equipment. If the ONTs receive one service (e.g., only video or only data), either the OLT or the video equipment is malfunctioning. If the ONTs have no services whatsoever, an OTDR can be used by a technician in the central office to see if there is an optical path discontinuity in the equipment patch cords, the fiber distribution rack, or the feeder fiber. Since the cabling within the CO is short, an optical pulse suppressor should be used with the OTDR (see Section 13.4) to locate a fault inside the CO. If these procedures do not locate the break point, it may be necessary to dispatch a maintenance truck to inspect the distribution cabinet.

If only a few ONTs are affected, the problem is in the distribution section of the FTTP network. As Figure 13.17 shows, the possibilities include damage at the distribution cabinet (zone 3), breaks in a distribution cable (zone 4) or a drop cable (zone 6), a fault at an access terminal (zone 5), or a malfunctioning regional power supply (zone 7). Here it usually is necessary for a troubleshooting technician to go to the site. Since the remainder of the network is operating properly, the use of an OTDR to look at the distribution and drop cables from the far end (point A in Figure 13.17) requires the instrument to be set at 1550 nm. This is necessary to avoid interference with the upstream 1310-nm traffic from the other ONTs.

A third case is when a single ONT is not working or cannot communicate properly with the OLT. If the ONT is not functioning at all, there may be a power outage

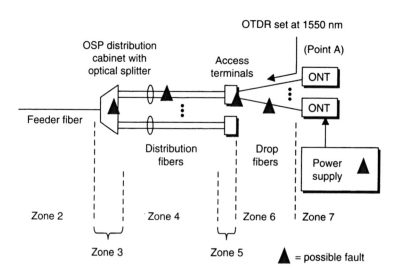

Figure 13.17. Possible fault locations in a scenario where only one or a few ONTs are affected by outages.

TABLE 13.4 Some ITU-T Recommendations for Optical Fiber Cable Maintenance Techniques

Recommendation Name	Description
L.40: "Optical fiber outside plant maintenance support, monitoring, and testing systems," Oct. 2000.	Describes minimal levels of maintenance and testing that are required to provide high reliability and quick response.
L.51: "Passive node elements for fiber optic networks," Apr. 2003.	Contains the general principles for generating performance requirements for passive optical nodes.
L.53: "Optical fiber maintenance criteria for access networks," Mar. 2003.	Describes the fundamental requirements, maintenance items, and testing methods for maintaining point-to-multipoint and ring optical networks.

or the ONT circuitry may be bad. When the ONT is functioning properly but can neither receive service nor communicate with the OLT, there is a fault in the fiber path between the ONT and the optical power splitter. If the communications between the ONT and the OLT are marginal, there may be a possible overly tight bend along the optical fiber line somewhere. In all cases, these malfunctions require that a service technician visits the customer site.

13.11.2 Troubleshooting Guidelines

The ITU-T has published a number of L-series recommendations for the maintenance of fiber optic cables in the outside plant. Table 13.4 lists some of these and states their application briefly.

13.12 SUMMARY

The deployment, operation, and maintenance of a PON require measurement techniques for verifying that the network has been configured properly and that its constituent components are functioning correctly. The main parameters that test instruments need to measure in an FTTP network are the optical power level at a variety of network points, loss of power as light passes through the various PON elements, and optical return loss.

The loss of individual fibers in the entire network should be checked before they are attached to the power splitter. This needs to be done bidirectionally and includes both the feeder fiber and the distribution and drop cables running between the splitter and the ONTs. The downstream measurements need to be done at 1490 and 1550 nm, since the attenuation varies with wavelength. Similarly, in the upstream direction the loss should be measured at 1310 nm.

The appropriate instruments include optical power meters, light sources emitting at the three specific FTTP wavelengths (1310, 1490, and 1550 nm), visual fault indicators, optical time-domain reflectometers, and special optical-loss test sets.

Optical power measurement is the most basic function in fiber optic metrology. However, this parameter is not a fixed quantity and can vary as a function of other parameters, such as time, distance along a link, wavelength, phase, and polarization. Therefore, two standard classes of power measurements in an optical system are the peak power and the average power. The *peak power* is the maximum power level in a pulse, which might be sustained for only a very short time. The *average power* is a measure of the power level averaged over a relatively long time period compared to the duration of an individual pulse.

An OTDR can give a detailed overall picture of the end-to-end link characteristics. This can be done from just one end. It includes parameters such as the attenuation of individual fiber segments; the location and losses of splices, connectors, optical splitters, and WDM couplers; and link anomalies. The anomalies can come from overly tight bends in the fiber (e.g., in equipment cabinets), possible cracks, and mismatches in fiber core size.

PROBLEMS

13.1 Using vendor data sheets, list the functions of several optical power meters that can be used for FTTP applications. Select and compare models from at least two different vendors. Consider functions such as dynamic range, wavelength range, capability to configure threshold settings, data storage features, size, weight, and battery use time.

13.2 Using vendor data sheets, list the functions of several instruments that can measure two-way optical return loss (ORL). Select and compare models from at least two different vendors. Consider functions such as test time, data storage features, configuration of threshold settings, power meter capabilities, size, weight, and battery use time.

13.3 (a) What percentage of light traveling in a fiber is reflected in the backward direction at a glass-to-air interface? Recall that $n_{air} = 1.00$ and $n_{glass} = 1.50$.
(b) What is the percent of light reflected if a gel with an index $n_{gel} = 1.30$ replaces the air interface?

13.4 Using vendor data sheets, compare several portable OTDRs that are optimized to have short dead zones. Consider factors such as event dead zone, attenuation dead zone, dynamic range, test time, test wavelengths, data storage capabilities, instrument interface features (e.g., does it have a touch screen?), size, weight, and battery use time.

13.5 If Rayleigh scattering is assumed to be uniform in all directions, show that the fraction S of scattered light that is trapped in the backward direction in an optical fiber is given by

$$S = \frac{\theta^2}{4}$$

where θ is the half-angle of the cone of captured rays. If θ = 0.12 radian, what fraction of the scattered light is captured by the fiber in the reverse direction?

13.6 Three 5-km-long fibers have been spliced together and an OTDR is used to measure the attenuation of the resultant fiber. If the OTDR displays the trace shown in Figure 13.18, what are the attenuations in decibels per kilometer of the three individual fibers? What are the splice losses in decibels? What are some possible reasons for the large splice loss occurring between the second and third fibers?

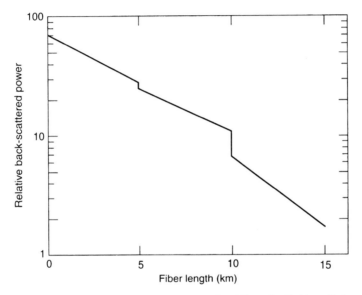

Figure 13.18. OTDR trace of three 5-km spliced fibers for Problem 13.6.

13.7 For light traveling in a fiber, let α be the attenuation of the forward-propagating light, β the attenuation of the backscattered light, and S the fraction of optical power scattered in the backward direction. The backscatter response of a rectangular pulse of width W from a point a distance L down the fiber as shown on the screen of the OTDR can be represented by the expression

$$P(L) = S\frac{\beta}{\alpha}P_0 e^{-2\alpha L}(1 - e^{-\alpha W})$$

where P_0 is the fiber input power. Show that for very short pulse widths the backscattered power is proportional to the pulse duration. (*Note:* This is the basis of operation of an OTDR.)

13.8 Using an OTDR, the uncertainty U of loss measurements as a function of the signal-to-noise ratio SNR can be approximated by $\log |U| = -0.2$ SNR +

0.6. Here U and SNR are given in decibels. Suppose that a 0.5-dB splice is located near the far end of a 20-km fiber that has a 1×8 optical splitter in the middle. What dynamic range must the OTDR have to measure the insertion loss of the splice event with an accuracy of $\pm 0.5\,\text{dB}$? Assume that the fiber attenuation is 0.25 dB and that the splitter insertion loss is 10.7 dB.

13.9 Show that when using an OTDR, an optical pulse width of 5 ns or less is required to locate a fault to within ± 0.5 m of its true position.

13.10 Verify the optical power losses for the mismatched fibers listed in Table 13.3.

REFERENCES AND FURTHER READING

1. D. Anderson, L. Johnson, and F. G. Bell, *Troubleshooting Optical Fiber Networks: Understanding and Using Optical Time-Domain Reflectometers*, Academic Press, San Diego, CA, 2004.
2. F. Caviglia, V. C. Di Biase, and A. Gnazzo, "Optical maintenance in PONs," *Opt. Fiber Technol.*, vol. 5, pp. 349–362, Oct. 1999.
3. S. Chabot and M. Leblanc, "Fiber optic testing challenges in point-to-multipoint PON testing," *EXFO Application Note 110*, www.exfo.com.
4. Bob Chomycz, *Fiber Optic Installer's Field Manual*, McGraw-Hill, New York, 2000.
5. D. Derickson, ed., contributors from Hewlett-Packard Co., *Fiber Optic Test and Measurement*, Prentice Hall, Upper Saddle River, NJ, 1998.
6. EXFO, *FTTx PON Guide*, 2nd ed., EXFO Electro-Optical Engineering, Quebec City, Canada, 2005.
7. C. Mas and P. Thiran, "A review on fault location methods and their application to optical networks," *Opt. Networks Mag.*, vol. 2, pp. 73–87, July–Aug. 2001.

14

NETWORK MANAGEMENT FUNCTIONS

In a traditional telecommunication system, a service provider is responsible for supplying and managing the transmission cable plant and the equipment on both ends of the network. The performance and operations management of this network requires the ability to configure and monitor network devices quickly and easily so that connections and services are always available. Early detection of actual or pending changes in the network performance status is critical in avoiding potential problems.

A driving force behind keeping the network fully operational and well tuned is the legal contract called a *service level agreement* (SLA). The terms of the SLA state that the service provider guarantees to their business customers a measurable *quality of service* (QoS). For example, an SLA may guarantee that service will be available 99.999 percent of the time with a designated bit error rate (BER) within a monthly or annual time period. If these guarantees are not met, the customer may receive a rebate or a one-time rate reduction. Thus, there is a financial incentive for the service provider to manage and monitor the key performance parameters of the network very closely.

In a PON the scenario is different from that for traditional telecommunication systems. Here the end users have the option to purchase their own premises equipment from any vendor on the open market. In this case the customers must maintain the communications gear in their premises, and the service provider is responsible only for the central office (CO) equipment and the outside cable plant up to the customer site. Since the service provider does not need to be involved with communications gear at the subscriber's premises, any failure or malfunction of this equipment is not considered to be a service interruption as it would be under the SLA terms in a traditional system.

In this chapter we examine some of the issues related to the *operation, administration, maintenance, and provisioning* (OAM&P) functions of a passive optical network.

FTTX Concepts and Applications, by Gerd Keiser
Copyright © 2006 John Wiley & Sons, Inc.

Note that in an actual system different groups of network operations personnel normally take separate responsibilities for issues such as administration aspects, performance monitoring, network integrity, access control, and information security. There is no special method of allocating the various management functions to particular groups of people, since each organization may take a different approach to fit its own needs. Here we look at these issues mainly from the point of view of the responsibilities of a service provider.

14.1 BASIC NETWORK MANAGEMENT

Figure 14.1 shows the components of a typical network management system and their relationships. The *network management console* is a workstation with specialized network management software. From such a console a human network manager can view the health and status of the network to verify that all devices are functioning properly, that they are configured correctly, and that their application software is up to date. A network manager can also see how the network is performing: for example, in terms of traffic loads and fault conditions in both the equipment and the cable plant. In addition, the console allows control of the network resources.

The *managed devices* are network components such as an optical line terminal (OLT), an optical network terminal (ONT), and backup batteries and power supplies.

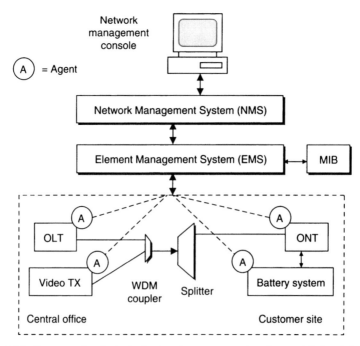

Figure 14.1. Components of a typical network management system and their relationships.

Each such device is monitored and controlled by its *element management system* (EMS). An important point to note here is that although typically the customer is responsible for the ONT premises equipment, to operate properly on a PON this equipment must be able to support status queries and control functions from the EMS.

As an example, ITU-T Rec. G.983.2 describes how the element management system of a BPON uses the *ONT management and control interface* (OMCI) to ensure proper operation of an ONT. This recommendation specifies the process for the exchange of management information between the OLT and ONT, and it covers the ONT management and control channel structure, the characteristics of the exchange protocol, and the format of detailed messages.

Management software modules, called *agents*, residing in a microprocessor within the network elements gather and compile information continuously on the status and performance of the devices being managed. The agents store this information in a *management information base* (MIB) at the central office and then provide this information to *management entities* within a *network management system* (NMS) that resides in the management workstation. A MIB is a logical base of information that defines data elements and their appropriate identifier, such as the fields in a database. This information may be stored in tables, counters, or switch settings. The MIB does not define how to collect or use data elements, but only specifies what the agent should collect and how to organize these data elements so that other systems can use them.

When agents notice problems in the element they are monitoring (e.g., reduction in optical power output from an OLT or an ONT, abnormal backup battery status, or excessive bit error rates), they send alerts to the management entities. Upon receiving an alert, the management entities can initiate one or more actions, such as operator notification, event logging, system shutdown, or automatic attempts at fault isolation or repair. The EMS also can query or poll the agents in the elements to check the status of certain conditions or variables. This polling can be automatic or operator-initiated.

14.2 MANAGEMENT FUNCTIONS

The International Standards Organization (ISO) has grouped network management functions into five generic categories: performance, configuration, accounting, fault, and security management. The principles for applying these functions to managing networks in general are described in the ITU-T Recommendation X.701, *System Management Overview*. In this section we define each of these categories and show how they relate to managing passive optical networks.

Related ITU Recommendations for optical systems include:

G.872: *Architecture of Optical Transport Networks* (Nov. 2001) describes the creation of the optical transport network (OTN). The term OTN refers to advanced DWDM-based networks that are capable of transporting a variety of heterogeneous signal types across optical backbones using multiple wavelengths. These networks use a special data formatting scheme called a *digital wrapper*.

G.874: *Management Aspects of the OTN Element* (Nov. 2001) addresses management functions for fault management, configuration management, and performance monitoring of the OTN elements.

G.874.1: *OTN Protocol-Neutral Management Information Model for the Network Element View* (Jan. 2002) provides a protocol-neutral management information model for managing network elements in the optical transport network, which can be used as the basis for defining protocol-specific management information models.

G.959.1: *Optical Transport Network Physical Layer Interfaces* (Dec. 2003) provides physical layer interface specifications for optical networks which may employ wavelength-division multiplexing in unidirectional, point-to-point, single-channel, and multichannel intra-office, short-haul, and long-haul systems.

G.7710: *Common Equipment Management Function Requirements* (Nov. 2001) addresses the equipment management functions inside an OTN element that are common to multiple technologies, for example, date and time, fault management, configuration management, account management, performance management and security management.

14.2.1 Performance Management

A telecommunication system will use the procedures of *performance management* to monitor and control key parameters that are essential to the proper operation of a network in order to guarantee a specific QoS to network users. In FTTP networks these parameters include a remote monitoring loop-back function, fault supervision, reporting of ONT failure statistics, and power switch off in case the ONT malfunctions and starts to flood the network with gibberish (see Section 14.3).

Examples of parameters that may be monitored at the physical level are bit error rate and optical power levels at both the OLT and the ONTs. The performance management procedure assigns threshold values to such parameters and informs the management system or generates alarms when these thresholds are reached.

14.2.2 Configuration Management

The goal of *configuration management* is to monitor both network setup information and network device configurations. The purpose of this is to track and manage the effects on network operation of the various constituent hardware and software elements. Configuration management allows a system to provision network resources and services, monitor and control their state, and collect status information. This provisioning may include setting optical power levels automatically (e.g., the output from an ONT in a GPON can be set at one of three different levels), assigning bandwidth or special features requested by a user, distributing software upgrades to agents, and reconfiguring equipment to isolate faults. Configuration management stores all this information in a readily accessible database so that when a problem occurs the database can be searched for assistance in solving the problem.

14.2.3 Accounting Management

The function of *accounting management* is to measure network-utilization parameters so that individuals or groups of users on the network can be regulated and billed for services appropriately. Thus, accounting management measures, collects, and records statistics on resource and network use. In addition, accounting management also may examine current usage patterns in order to allocate network usage quotas. From the statistics gathered, the service provider can then generate a bill or a tariff for use of the service.

14.2.4 Fault Management

Faults in a network, such as physical cuts in a fiber transmission line or failure of an OLT or an ONT can cause the entire FTTP network or portions of it to be inoperable. Since network faults can result in system downtime or unacceptable network degradation, *fault management* is one of the most widely implemented and important network management functions. With the growing dependence of people on network resources for carrying out their work and communications, users expect rapid and reliable resolutions of network fault conditions. As Figure 14.2 illustrates, fault management involves the following processes:

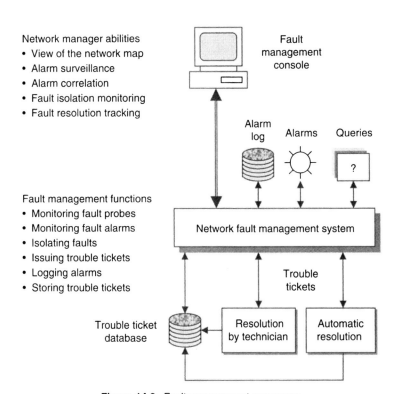

Figure 14.2. Fault management processes.

1. *Alarm surveillance* is used to report alarms and their possible causes to the NMS. These alarms are activated in response to detection of fault or degradation symptoms and may have different levels of severity. Fault management also provides a summary of unresolved alarms and allows the network manager to retrieve and view the alarm information from an alarm log.
2. *Fault-isolation* techniques determine the origin, location, and possible cause of faults either automatically or through the intervention of a network manager. This can include functions such as alarm correlation from different parts of the network and diagnostic testing.
3. *Trouble tickets* are issued by the NMS once the faults are isolated. These tickets indicate what the problem is and possible means of how to resolve the fault. When they are issued the tickets go to either a technician for manual intervention or an automatic fault-correction mechanism. When the fault or degradation is corrected, this fact and the resolution method are indicated on the trouble ticket, which is then stored in a database.
4. *Operational testing* is performed once the problem has been fixed. In this procedure the repair is operationally tested on all major subsystems of the network. It involves requesting performance tests, tracking the progress of these tests, and recording the results.

An important factor in troubleshooting faults is to have a comprehensive physical and logical map of the network. Ideally, this map should be part of a software-based management system that can show the network connectivity and the operational status of the constituent elements of the network on a display screen. With such a map, failed or degraded devices can be viewed easily and corrective action can be taken immediately. This is especially important for large networks of PONs.

14.2.5 Security Management

The concept of security covers a wide range of disciplines. These include applying encryption techniques to certain types of traffic, setting up virus-protection software, establishing access authentication procedures, implementing special firewall software to prevent unauthorized access of corporate information, and developing security policies and principles. The principal goal of network *security management* is to establish and enforce guidelines to control access to network resources. This control is needed to prevent intentional or unintentional sabotage of network capabilities and to prevent viewing or modification of sensitive information by people who do not have appropriate access authorization.

In the case of PON architectures, since the downstream data from the OLT is broadcast to all ONTs, every message transmitted can be seem by all the users attached to the PON. Each of the BPON, EPON, and GPON architectures follow a different approach to ensure that users are allowed to access only the data intended for them. A standard technique is to use some type of *encryption*, whereby data are transformed into an unintelligible format at the sending end to protect them against unauthorized disclosure, modification, utilization, or destruction as they travel through the network.

The ITU-T G.983 BPON standard prescribes *churning* as a security mechanism for traffic encryption in the downstream direction only. This mechanism uses a simple *substitution cipher* in which the upper and lower 4 bits of each byte are encrypted using separate keys. Each churning key is set and updated at least once every second by individual ONTs and are sent upstream for use by the OLT. An OLT also may request passwords from a particular ONT to prevent access by a malicious user pretending to be another ONT. Upstream transmissions are not encrypted based on the assumption that eavesdropping on an upstream link is difficult.

The Ethernet protocol does not have any built-in security mechanism. However, developers of EPON equipment are incorporating point-to-multipoint traffic security mechanisms into their OLT and ONT offerings. Many of these are based on standard multilayered security mechanisms for IP traffic, such as firewalls, virtual private network technology, Internet protocol security, and tunneling.

The GPON standard describes the use of a point-to-point encryption mechanism. This is the *Advanced Encryption Standard* (AES), which is used to protect the information payload of the data field in the GPON frame. The AES encryption algorithm encrypts and decrypts 128-bit data blocks from their original format called *plaintext* to an unintelligible form called *ciphertext*. The cipher keys can have lengths of 128, 192, or 256 bits, which makes the encryption extremely difficult to compromise. Decrypting the ciphertext converts the data back into its original form.

14.3 OAM&P IN FTTP NETWORKS

To satisfy customer demands and expectations for high-quality triple-play services, FTTP networks must employ reliable and efficient OAM&P procedures. These procedures support functions such as billing, security, maintenance, provisioning, and the monitoring of network performance. This can be achieved through the use of standard or enhanced *operation support system* (OSS) software tools in the NMS console. Numerous OSS programs are available with colorful *graphical user interfaces* (GUIs) that can be enabled through a Web-based browser, automatic service-provisioning capabilities, customizable reporting of all kinds, and a wide selection of applications that allow the network manager to configure and control hundreds of elements easily.

Provisioning deals with providing and configuring various grades and types of voice, data, and video services to a customer. For this either a human operator or an automatic mechanism needs to determine if the equipment at the customer's premises can accommodate the services requested. For example, provisioning must determine if the ONT has the capability to handle a specific data rate and whether or not it has integrated testing capabilities that can support the remote monitoring and control functions required in the SLA. This is particularly important to the service provider, since remote management and determination of problems in an ONT can

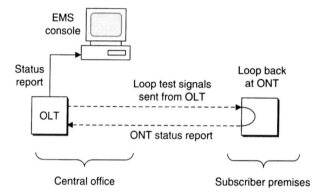

Figure 14.3. Remote loop-back testing using status-monitoring signals initiated by an OLT.

preclude the need for a costly service call. Note that the telecommunications industry refers to such service calls made by a technician as *truck rolls*.

A basic part of the *maintenance* (or fault management) function is to ensure that the SLA parameters are being met. One part of this activity is to carry out proactive preventive maintenance to avoid potential faults or degradations in all parts of the FTTP network. If faults do occur, the maintenance process needs to locate and clear them as quickly as possible to maintain customer satisfaction and to meet the SLA specifications.

The integrated testing capabilities for exchanging status information between an ONT and an OLT include a remote monitoring loop-back function, detection of electronics or cable plant faults, reporting of ONT failure statistics, and power shutdown in case the ONT malfunctions and starts to flood the network with gibberish. As shown in Figure 14.3, during remote *loop-back testing* status-monitoring signals are initiated by an OLT and sent to an ONT. From there ONT status reports are returned to the central office. These reports can indicate that everything is working well or they can be used to determine possible causes of high bit error rates or difficulties with specific traffic patterns.

14.4 SUMMARY

A driving force behind keeping a network fully operational and well tuned is the legal contract called a service level agreement (SLA). The terms of the SLA state that the service provider guarantees to its business customers a measurable quality of service. In a PON the end users have the option to purchase their own premises equipment from any vendor on the open market and must maintain the communications gear in their premises. Since the service provider is responsible only for the central office equipment and the outside cable plant up to the customer site, any failure or malfunction of this equipment is not considered to be a service interruption as it would be under the SLA terms in a traditional system. Nevertheless, a service provider needs to be both proactive and reactive in solving pending or real PON faults, and in many cases when there are malfunctions at an ONT the service provider needs to support the customer in resolving the faults.

A typical network management system has a network management workstation with specialized network management software. From such a workstation a human network manager can view the health and status of the network to verify that all devices are functioning properly, that they are configured correctly, and that their application software is up to date. A network manager also can see how the network is performing, for example, in terms of traffic loads and fault conditions in both the equipment and the cable plant. In addition, the console allows control of network resources.

Network management functions can be grouped into the five generic categories of performance, configuration, accounting, fault, and security management. The principles for applying these functions to managing networks in general are described in ITU-T Recommendation X.701, *System Management Overview*.

Faults in a network, such as physical cuts in a fiber transmission line or failure of an OLT or an ONT, can cause the entire FTTP network or portions of it to be inoperable. Since network faults can result in system downtime or unacceptable network degradation, fault management is one of the most widely implemented and important network management functions. With the growing dependence of people on network resources for carrying out their work and communications, users expect rapid and reliable resolutions of network fault conditions.

PROBLEMS

14.1 List and describe in a few sentences at least 10 factors that need to be considered when establishing a service level agreement between a telecommunications carrier and a large enterprise.

14.2 List and discuss the importance of at least four parameters of FTTP networks that software agents in an element management system might monitor and store in a management information base. Include several that are routine status parameters and several that would trigger an alarm or warning indicator if they change significantly or do not meet specified thresholds.

14.3 List and discuss some configuration management functions that may take place in FTTP networks. Consider one set of functions for a network that serves only home users and another set of functions that are more applicable to FTTP networks serving businesses and other large organizations.

14.4 Consider the fault management process shown in Figure 14.2. List and discuss the specific fault-management steps and actions that might take place when the following failures and degradations occur:

(**a**) the failure of one ONT in the network;

(**b**) a slowly degrading fiber optic connector on a drop cable;

(**c**) a cut in an eight-fiber distribution cable.

FURTHER READING

1. Federal Information Processing Standards (FIPS) Publication 197, "Advanced Encryption Standard," Nov. 26, 2001.
2. J. D. Huh et al. "Key management device and method for providing security service in Ethernet-based passive optical network," U.S. patent application 20050008158, Jan. 13, 2005.
3. ITU-T Recommendation G.983.2, "ONT Management and Control Interface Specification for BPON," June 2002.
4. G. Keiser, *Local Area Networks*, 2nd ed., McGraw-Hill, Burr Ridge, IL, 2002.
5. S. Thomas and D. Wagner, "Insecurity in ATM-based passive optical networks," *IEEE International Conf. Communications (ICC 2002) Proc., Optical Networking Symp.*

APPENDIX A

UNITS, PHYSICAL CONSTANTS, AND CONVERSION FACTORS

A.1 INTERNATIONAL SYSTEM OF UNITS

Parameter	Unit	Symbol	Dimension
Length	meter	m	
Mass	kilogram	kg	
Time	second	s	
Temperature	kelvin	K	
Electric current	ampere	A	
Frequency	hertz	Hz	1/s
Force	newton	N	$kg \cdot m/s^2$
Pressure	pascal	Pa	N/m^2
Energy	joule	J	$N \cdot m$
Power	watt	W	J/s
Electric charge	coulomb	C	$A \cdot s$
Potential	volt	V	J/C
Conductance	siemens	S	A/V
Resistance	ohm	Ω	V/A

FTTX Concepts and Applications, by Gerd Keiser
Copyright © 2006 John Wiley & Sons, Inc.

A.2 PHYSICAL CONSTANTS

Constant	Symbol	Value (mks units)
Speed of light in vacuum	c	2.99793×10^8 m/s
Electron charge	q	1.60218×10^{-19} C
Planck's constant	h	6.6256×10^{-34} J·s
Boltzmann's constant	k_B	1.38054×10^{-23} J/K
k_B/q at $T = 300$ K	—	0.02586 eV
Electron volt	eV	1.60218×10^{-19} J
Base of natural logarithms	e	2.71828
Pi	π	3.14159

A.3 CONVERSION FACTORS

From	To	Conversion
cm	inch	cm × 0.3937
km	mile	km × 0.6214
inch	cm	inch × 2.5400
mile	km	mile × 1.6093
eV	J	eV × 1.60218×10^{-19}
radian	degree	radian × 57.296
°C	°F	°C × 1.8 + 32
K	°C	K + 273

APPENDIX B

ACRONYMS

ABR	Available bit rate
ADSL	Asymmetric DSL
AES	Advanced Encryption Standard
AM-VSB	Amplitude-modulated vestigial sideband
ANSI	American National Standards Institute
A-NZDSF	Advanced nonzero dispersion-shifted fiber
APC	Angle-polished connector
APD	Avalanche photodiode
APON	ATM PON
ASE	Amplified spontaneous emission
ASK	Amplitude shift keying
ATM	Asynchronous transfer mode
AWG	Arrayed waveguide grating
BER	Bit error rate
BIP	Bit interleaved parity
B-ISDN	Broadband ISDN
BPON	Broadband PON
CATV	Cable television
CBR	Constant bit rate
CD	Chromatic dispersion
CDV	Cell delay variation
CER	Cell error rate
CLR	Cell loss ratio
CNR	Carrier-to-noise ratio
CO	Central office
CRC	Cyclic redundancy check
CSO	Composite second order

FTTX Concepts and Applications, by Gerd Keiser
Copyright © 2006 John Wiley & Sons, Inc.

CTB	Composite triple beat
CTD	Cell transfer delay
CWDM	Coarse WDM
DA	Destination address
DBA	Dynamic bandwidth allocation
DBR	Dynamic bandwidth report
DFB	Distributed feedback (laser)
DMT	Discrete multitone
DSF	Dispersion-shifted fiber
DSL	Digital subscriber line
DUT	Device under test
DWDM	Dense WDM
EDFA	Erbium-doped fiber amplifier
EFM	Ethernet in the first mile
EIA	Electronic Industries Alliance
EM	Electromagnetic
EMF	Equipment management function
EMS	Element management system
EO	Electrooptical
EPON	Ethernet PON
ER	Extinction ratio
FBG	Fiber Bragg grating
FCS	Frame check sequence
FDF	Fiber distribution frame
FEC	Forward error correction
FOTP	Fiber-optic test procedure
FP	Fabry–Perot
FRPE	Flame-retardant polyethylene
FSAN	Full Service Access Network (committee)
FTTB	Fiber-to-the-business
FTTC	Fiber-to-the-curb
FTTH	Fiber-to-the-home
FTTN	Fiber-to-the-neighborhood
FTTO	Fiber-to-the-office
FTTP	Fiber-to-the-premises
FTTU	Fiber-to-the-user
FTTX	Fiber-to-the-x
FWHM	Full-width half-maximum
FWM	Four-wave mixing
GEM	GPON encapsulation method
GigE	Gigabit Ethernet
GPM	GPON physical medium dependent
GPON	Gigabit PON
GSR	GPON service requirements

GTC	GPON transmission convergence
GUI	Graphical user interface
HASB	High air speed blown
HDTV	High-definition television
HFC	Hybrid-fiber coax
IEC	International Electrotechnical Commission
IEEE	Institute for Electrical and Electronic Engineers
IP	Internet protocol
IPSec	Internet protocol security
ISDN	Integrated Services Digital Network
ISO	International Standards Organization
ISP	Internet service provider
ITU	International Telecommunications Union
ITU-T	Telecommunications Sector of the ITU
LAN	Local area network
LC	Lucent (connector)
LCP	Local convergence point
LED	Light-emitting diode
LLID	Logical link identifier
MAC	Media access control
MAN	Metropolitan area network
MCR	Minimum cell rate
MDU	Multiple dwelling unit
MHU	Multiple hospitality unit
MIB	Management information base
MPCP	Multipoint control protocol
MPEG	Moving Pictures Experts Group
MTBF	Mean time between failures
MT-RJ	Media termination—recommended jack (connector)
MTU	Multiple tenant unit
MU	Miniature unit (connector)
MZI	Mach–Zehnder interferometer
NA	Numerical aperture
NAP	Network access point
NEC	National Electrical Code
NIC	Network interface card
NIU	Network interface unit
NMS	Network management system
nrt-VBR	Non-real-time variable bit rate
NRZ	Non-return-to-zero
NTT	Nippon Telegraph and Telephone
NZDSF	Nonzero dispersion-shifted fiber
OAM	Operations, administration, and maintenance
OAM&P	Operation, administration, maintenance, and provisioning

OC-N	Optical carrier—level N
ODN	Optical data network
OFC	Optical fiber conductive
OFCP	Optical fiber conductive plenum
OFCR	Optical fiber conductive riser
OFN	Optical fiber nonconductive
OFNP	Optical fiber nonconductive plenum
OFNR	Optical fiber nonconductive riser
OLT	Optical line terminal
OLTS	Optical-loss test set
OMCI	ONT management and control interface
ONT	Optical network terminal
ONU	Optical network unit
OPS	Optical pulse suppressor
ORL	Optical return loss
OSA	Optical spectrum analyzer
OSI	Open System Interconnect
OSNR	Optical signal-to-noise ratio
OSP	Outside cable plant
OSS	Operations support system
OTDR	Optical time-domain reflectometer
OTN	Optical transport network
P2P	Point-to-point
PCR	Peak cell rate
PDH	Plesiochronous digital hierarchy
PDU	Protocol data unit
PE	Polyethylene
PHY	Physical layer
PLC	Planar lightwave circuit
PLI	Payload length indicator
PLO	Physical layer overhead
PLOAM	Physical layer OAM
PLS	Power leveling sequence
PMD	Physical media dependent
PMD	Polarization mode dispersion
PON	Passive optical network
POP	Point of presence
PSTN	Public switched telephone network
PTI	Payload-type indicator
PU	Polyurethane
PVC	Polyvinyl chloride
QAM	Quadrature amplitude modulation
QoS	Quality of service
QPSK	Quadrature phase-shift keying

RF	Radio frequency
RIN	Relative intensity noise
RS	Reed–Solomon code
RTT	Round-trip time
rt-VBR	Real-time variable bit rate
RZ	Return-to-zero
SA	Source address
SAN	Storage area network
SBS	Stimulated Brillouin scattering
SC	Subscriber connector *or* Square connector
SCR	Sustainable cell rate
SDH	Synchronous digital hierarchy
SDU	Single dwelling unit
SFD	Start frame delimiter
SFF	Small form factor
SFP	SFF pluggable
SFU	Single-family unit
SLA	Service level agreement
SNMP	Simple network management protocol
SNR	Signal-to-noise ratio
SOA	Semiconductor optical amplifier
SOHO	Small home or small office
SONET	Synchronous optical network
SPD	Start-of-packet delimiter
SPE	Synchronous payload envelope
SPM	Self-phase modulation
SRS	Stimulated Raman scattering
ST	Straight tip (connector)
STM-1	Synchronous transport module—level 1
STS-1	Synchronous transport signal—level 1
TC	Transmission control
T-CONT	Traffic container
TCP	Transmission control protocol
TDM	Time-division multiplexing
TDMA	Time-division multiple access
TE	Transverse electric
TEC	Thermoelectric cooler
TFF	Thin-film filter
TIA	Telecommunication Industry Association
UBR	Unspecified bit rate
UL	Underwriters' Laboratories
UPS	Uninterruptible power supply
US BW	Upstream bandwidth
VC	Virtual channel

VFL	Visual fault locator
VLAN	Virtual LAN
VOD	Video on demand
VP	Virtual path
VPN	Virtual private network
WAN	Wide area network
WDM	Wavelength-division multiplexing
Wi-Fi	Wireless fidelity
WiMax	Worldwide interoperability for microwave access
XPM	Cross-phase modulation

APPENDIX C

VIDEO TRANSMISSION

A key application of a passive optical network is to use a separate wavelength for analog video transmission. There are three basic metrics for determining the fidelity of such signals:

- *Carrier-to-noise ratio* (CNR) measures the level of the video carrier signal to the noise. If this parameter is too low, the viewer will see a snow of white spots on the screen.
- *Composite second order* (CSO) is a signal distortion that occurs when one or more signal carriers experiences a second-order nonlinear effect. For example, when two carriers interact to produce a side tone that is at the same frequency as the signal desired. It is measured in decibels.
- *Composite triple beat* (CTB) is a signal distortion that occurs when one or more signal carriers experiences a third-order nonlinear effect. For example, when three carriers interact to produce a side tone that is at the same frequency as the desired signal. It is measured in decibels.

CSO and CTB appear as lines and patterns on the screen. To have a high-quality video signal, the minimum standards are a CNR greater than or equal to 43 dB, a CSO of less than or equal to -51 dBc (decibels relative to the carrier), and a CTB of less than or equal to -51 dBc.

To maintain such high signal levels at an ONT, a passive optical network makes use of one or more stages of optical amplification at the OLT in order to boost the optical signal level to an appropriately high value. A typical output from an erbium-doped fiber amplifier (EDFA) used in the central office is around 20 dBm.

FTTX Concepts and Applications, by Gerd Keiser
Copyright © 2006 John Wiley & Sons, Inc.

APPENDIX D

COMMUNICATIONS SIGNALS

In designing a communication link, an important consideration is the format of the digital signal that is transmitted. This is significant because the receiver must be able to extract precise *timing information* from the incoming signal. The three main purposes of *timing* are:

- To allow the signal to be sampled by the receiver at the time the signal-to-noise ratio is a maximum
- To maintain proper spacing between pulses
- To indicate the start and end of each timing interval

In addition, since errors resulting from channel noise and distortion mechanisms can occur in the signal-detection process, it may be desirable for the signal to have an inherent error-detecting capability, as well as an error-correction mechanism, if it is needed or is practical. These features can be incorporated into the data stream by structuring or *encoding* the signal. Generally, one does this by introducing extra bits into the raw data stream at the transmitter on a regular and logical basis and extracting them again at the receiver. This process is called *channel coding* or *line coding*. Some examples of generic encoding techniques are presented in this section.

Error detection is accomplished by means of a standard cyclic redundancy check. This technique is described in Section D.3.

D.1 NRZ AND RZ SIGNAL FORMATS

The simplest method for encoding data is the unipolar *non-return-to-zero (NRZ)* code. *Unipolar* means that a logic 1 is represented by a voltage or light pulse that fills an entire bit period, whereas for a logic 0 no pulse is transmitted, as shown in Figure D.1. The coded patterns in this figure are for the data sequence 1010110. If

---

FTTX Concepts and Applications, by Gerd Keiser
Copyright © 2006 John Wiley & Sons, Inc.

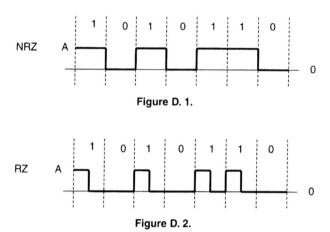

Figure D. 1.

Figure D. 2.

1 and 0 pulses occur with equal probability and if the amplitude of the voltage pulse is A, the average transmitted power for this code is $A^2/2$. In optical systems one typically describes a pulse in terms of its optical power level. In this case the average power for an equal number of 1 and 0 pulses is $P/2$, where P is the peak power in a 1 pulse. An NRZ code is simple to generate and decode, but it possesses no inherent error-monitoring or error-correcting capabilities, and it contains no timing features.

The lack of timing capabilities in an NRZ code can lead to misinterpretations of the bit stream at the receiver. For example, since there are no level transitions from which to extract timing information in a long sequence of NRZ 1's or 0's, a long string of N identical bits could be interpreted as either $N+1$ or $N-1$ bits unless highly stable (and expensive) clocks are used. This problem can be alleviated with a code that has transitions at the beginning of each bit interval when a binary 1 is transmitted and no transition for a binary 0. This can be achieved with a *return-to-zero* (RZ) code, as shown in Figure D.2. Here the pulse for a 1 bit occupies only the first half of the bit interval and returns to zero in the second half of the bit interval. No pulse is used for a 0 bit.

D.2 BLOCK CODES

Introducing *redundant bits* into a data stream can be used to provide adequate timing and to have error-monitoring features. A popular and efficient encoding method for this is the class of *mBnB block codes*. In this class of codes, blocks of m binary bits are converted to longer blocks of $n > m$ binary bits. As a result of the additional redundant bits, the required bandwidth increases by the ratio n/m. For example, in an *mBnB* code with $m=1$ and $n=2$, a binary 1 is mapped into the binary pair 10, and a binary 0 becomes 01. The overhead for such a code is 50 percent.

Suitable *mBnB* codes for high data rates are the 3B4B, 4B5B, 5B6B, and 8B10B codes. If simplicity of the encoder and decoder circuits is the main criterion, the 3B4B format is the most convenient code. The 5B6B code is the most advantageous if bandwidth reduction is the major concern. Various versions of Ethernet use the 3B4B, 4B5B, or 8B10B formats.

D.3 CYCLIC REDUNDANCY CHECK BASICS

The *cyclic redundancy check* (CRC) technique is based on a binary division process involving the data portion of a packet and a sequence of redundant bits. The basic procedure is as follows:

Step 1. At the sender end a string of n zeros is added to the data unit on which error detection will be performed. For example, this data unit may be a packet. The characteristic of the redundant bits is such that the result (packet plus redundant bits) is exactly divisible by a second predetermined binary number.

Step 2. The new enlarged data unit is divided by the predetermined divisor using binary division. If the number of bits added to the data unit is n, the number of bits in the predetermined divisor is $n + 1$. The remainder that results from this division is called the *CRC remainder* or simply the CRC. The number of digits in this remainder is equal to n. For example, if $n = 3$, it may be the binary number 101. Note that the remainder might also be 000 if the two numbers are exactly divisible.

Step 3. The n zeros that were added to the data unit in step 1 are replaced by the n-bit CRC. The composite data unit is then sent through the transmission channel.

Step 4. When the data unit plus the appended CRC arrives at the destination, the receiver divides this incoming composite unit by the same divisor that was used to generate the CRC.

Step 5. If there is no remainder after this division occurs, it is assumed that there are no errors in the data unit and it is accepted by the receiver. A remainder indicates that some bits became corrupted during the transmission process, and therefore the data unit is rejected.

Instead of using a string of 1 and 0 bits, the CRC generator is normally represented by an algebraic polynomial with binary coefficients. The advantage of using a polynomial is that it is simple to visualize and to perform the division mathematically. Commonly used polynomials and their binary equivalents for CRC generation are designated as CRC-8, CRC-10, CRC-16, and CRC-32. The numbers 8, 10, 16, and 32 refer to the size of the CRC remainder. Thus, the CRC divisors for these polynomials are 9, 11, 17, and 33 bits, respectively. The first two polynomials are used in ATM networks, whereas CRC-32 is used in IEEE 802 LANs. CRC-16 is used in bit-oriented protocols, such as the High-level Data Link Control (HDLC) standard, where frames are viewed as a collection of bits.

Example. The generator polynomial $x^7 + x^5 + x^2 + x + 1$ can be written as

$$1 \times x^7 + 0 \times x^6 + 1 \times x^5 + 0 \times x^4 + 0 \times x^3 + 1 \times x^2 + 1 \times x^1 + 1 \times x^0$$

where the exponents on the variable x represent bit positions in a binary number and the coefficients correspond to the binary digits at these positions. Thus, the generator polynomial given here corresponds to the 8-bit binary representation 10100111.

A polynomial needs to have the following properties:

- It should not be divisible by x. This condition guarantees that the CRC can detect all burst errors that have a length less than or equal to the degree of the polynomial.
- It should be divisible by $x + 1$. This allows the CRC to detect all bursts that affect an odd number of bits.
- Given these two rules, the CRC can also detect with a probability

$$P_{ed} = 1 - \frac{1}{2^N} \tag{D.1}$$

any burst errors that have a length greater than the degree N of the generator polynomial.

APPENDIX E

TELCORDIA GENERIC REQUIREMENTS FOR PON APPLICATIONS

Telcordia Technologies provides a wide range of generic requirements for telecommunications network components and equipment. Listed below are some of the Telcordia Generic Requirements for FTTP applications.

- GR-13: *Generic Requirements for Pedestal Terminal Closures.* This document includes requirements for closures used in passive optical networks.
- GR-3120: *Generic Requirements for Hardened Fiber Optic Connectors.* This document includes requirements for connectors that are environmentally hardened and can be mated in the field.
- GR-3121: *Generic Requirements for Below Ground Cabinets.* This includes cabinets intended to house passive optical network components. The cabinets could be located in a service provider right-of-way or on a customer's premises, such as an apartment building, and that can serve as a distribution hub for FTTP services.
- GR-3122: *Generic Requirements for FITS (Factory Installed Termination System).* These requirements include field-deployed products that have been assembled in the factory. Normally, products such as cables, cable assemblies, cross-connect boxes, enclosures, and fiber distribution hubs are shipped to a work site and assembled in the field. Based on GR-3122, a FITS product would be a complete distribution and feeder cabling system that is shipped directly to the field from the factory.
- GR-3123: *Generic Requirements for Indoor Fiber Distribution Hubs.* This document includes requirements for enclosures that house PON components and that could be located on a customer premises, such as an apartment building, and that can serve as a distribution hub for FTTP services.

FTTX Concepts and Applications, by Gerd Keiser
Copyright © 2006 John Wiley & Sons, Inc.